P9-APW-937

Siegel · Lectures on the Geometry of Numbers

Carl Ludwig Siegel

Lectures on the
Geometry of Numbers

Notes by B. Friedman

Rewritten by
Komaravolu Chandrasekharan
with the Assistance of Rudolf Suter

With 35 Figures

No Longer Property of
Phillips Memorial Library

Springer-Verlag
Berlin Heidelberg New York
London Paris Tokyo
Hong Kong

Phillips Memorial
Library
Providence College

Komaravolu Chandrasekharan
Mathematik, ETH Zürich
CH-8092 Zürich, Switzerland

QA
241.5
S54
1989
✓

Mathematics Subject Classification (1980):
01-XX, 11-XX, 12-XX, 15-XX, 20-XX, 32-XX, 51-XX, 52-XX

ISBN 3-540-50629-2 Springer-Verlag Berlin Heidelberg New York
ISBN 0-387-50629-2 Springer-Verlag New York Berlin Heidelberg

Library of Congress Cataloging-in-Publication Data.
Siegel, Carl Ludwig, 1896–1981
Lectures on the geometry of numbers / Carl Ludwig Siegel; notes by B. Friedman; rewritten
by Komaravolu Chandrasekharan with the assistance of Rudolf Suter. p. cm.
Bibliography: p. Includes index.
ISBN 0-387-50629-2 (U.S. : alk. paper)
1. Geometry of numbers. I. Chandrasekharan, K. (Komaravolu), 1920 –.
II. Suter, Rudolf, 1963 –. III. Title. QA241.5.S54 1989 512'.5 – dc19 89–5946 CIP

This work is subject to copyright. All rights are reserved, whether the whole or part of the material
is concerned, specifically those of translation, reprinting, reuse of illustrations, recitation, broad-
casting, reproduction on microfilms or in other ways, and storage in data banks. Duplication of this
publication or parts thereof is only permitted under the provisions of the German Copyright Law
of September 9, 1965, in its version of June 24, 1985, and a copyright fee must always be paid. Viola-
tions fall under the prosecution act of the German Copyright Law.

© Springer-Verlag Berlin Heidelberg 1989
Printed in the United States of America

2141/3140-543210   Printed on acid-free paper

Preface

Carl Ludwig Siegel lectured on the *Geometry of Numbers* at New York University during 1945–46. There were hardly any books on the subject at that time other than Minkowski's original one. The freshness of his approach still lingers, and his presentation has its attractions for the aspiring young mathematician.

When he received a request, many years later, for permission to reissue the notes of those lectures, transcribed by the late B. Friedman, he declined. Those who know of Siegel's insistence on the expurgation of the first printing of his lecture notes on *Transcendental Numbers*, or of his insistence on the dissociation of his name from a projected English translation of his lecture notes on *Analytische Zahlentheorie* (Göttingen, 1963/64), which had therefore to be abandoned, need hardly be told that he had his own requirements.

We had occasion to discuss the matter further, and it was agreed that the notes should be published only after they had been checked, corrected, and rewritten, and he handed over his personal copy to me. I found that the task of revision required far more attention to detail than I had at first glance thought necessary. Other avocations prevented me from completing the work until the summer of 1987, when it happened that Rudolf Suter got actively interested in what I had been trying to carry out. His helpful and critical comments came to me as an unexpected stimulus, and the present version is the result. It is a pleasure for me to acknowledge Suter's assistance.

Admirers of Siegel's style can scarcely fail to notice his uncanny skill, and perspicuity in argument, with an occasional flash of wit, as they progress with the reading.

E.T.H. Zürich

31 March 1988   K. Chandrasekharan

Table of Contents

Chapter I
Minkowski's Two Theorems

Chapter II
Linear Inequalities

Chapter III
Theory of Reduction

Chapter I

Minkowski's Two Theorems

Lectures I to IV

Lecture I

§1. Convex sets

Consider an n-dimensional real Euclidean space $\mathbb{R}^n$, $n \geq 1$. Assume that a rectangular coordinate system with origin at some point O is set up in $\mathbb{R}^n$, so that the coordinates of any point $P \in \mathbb{R}^n$ are $x_1, \ldots, x_n$. For simplicity we shall represent the point P by the vector $x = (x_1, \ldots, x_n)$. The origin O is then represented by the zero-vector $0 = (0, \ldots, 0)$.

A non-empty set $\mathcal{L}$ contained in $\mathbb{R}^n$ is called a linear manifold in $\mathbb{R}^n$, if whenever any two different points P and Q belong to $\mathcal{L}$, the infinite straight line passing through P and Q belongs to $\mathcal{L}$. Analytically the definition can be formulated as follows:

Let x be the vector associated with P, and y the vector associated with Q. Then $\mathcal{L}$ is a linear manifold, if whenever it contains P and Q, it contains every point represented by a vector of the form $\lambda x + \mu y$, where λ, μ are arbitrary real numbers, such that $\lambda + \mu = 1$.

Any linear manifold $\mathcal{L}$ has a dimension m, which is an integer not greater than n, and which can be found as follows.

Let P_0 be a point in $\mathcal{L}$. If $\mathcal{L}$ contains no other point, then $m = 0$. Otherwise let P_1 be another point in $\mathcal{L}$. Then all points on the line passing through P_0, P_1 belong to $\mathcal{L}$. If $\mathcal{L}$ contains no point besides those on the line through P_0 and P_1, then $m = 1$. Otherwise let P_2 be a point in $\mathcal{L}$ which is not on the line through P_0, P_1. Then all points of the plane determined by the points P_0, P_1, P_2 belong to $\mathcal{L}$. If $\mathcal{L}$ contains no point outside this plane, then $m = 2$. Otherwise let P_3 be a point in $\mathcal{L}$ outside the plane through P_0, P_1, P_2, and we can continue this procedure. Since the highest possible value of m is n, it is clear that this procedure terminates and gives a definite value of m. Note that $\mathcal{L}$ is completely determined by the m-dimensional tetrahedron, or m-simplex, $P_0 P_1 \ldots P_m$ obtained in the course of the proof. Analytically this means the following.

If $x^{(i)}$ is the vector representing the point P_i, then a point P belongs to $\mathcal{L}$ if and only if its vector can be written as $\lambda_0 x^{(0)} + \lambda_1 x^{(1)} + \ldots + \lambda_m x^{(m)}$, where $\lambda_0, \lambda_1, \ldots, \lambda_m$ are arbitrary real numbers, such that $\lambda_0 + \lambda_1 + \ldots + \lambda_m = 1$.

A non-empty set $\mathcal{K} \subset \mathbb{R}^n$ is called a *convex set* if whenever P and Q belong to $\mathcal{K}$, the *segment* joining P and Q belongs to $\mathcal{K}$. Analytically the definition can be formulated in this way: if P is represented by the vector x, and Q by the vector y, then $\mathcal{K}$ is a convex set if with P and Q it contains also every point with a vector of the form $\lambda x + \mu y$, where $\lambda \geq 0$, $\mu \geq 0$, and $\lambda + \mu = 1$.

Just as before, we can find a number m, such that $\mathcal{K}$ is contained in a linear manifold $\mathcal{L}_m$ of dimension m but not contained in any $\mathcal{L}_r$ for $r < m$.

Let P_0 be a point in $\mathcal{K}$. If $\mathcal{K}$ contains no other point, then $m = 0$. Otherwise let P_1 be another point in $\mathcal{K}$; then all points in the segment $P_0 P_1$ belong to $\mathcal{K}$. If $\mathcal{K}$ is contained in the infinite straight line passing through P_0 and P_1, then $m = 1$, and so on. We can thus find an m-dimensional tetrahedron, or m-simplex, $P_0 P_1 \ldots P_m$, all of whose points belong to $\mathcal{K}$.

If $x^{(i)}$ is the vector representing the vertex P_i of the simplex, then a point P belongs to $\mathcal{K}$, if its vector can be written as $\lambda_0 x^{(0)} + \lambda_1 x^{(1)} + \ldots + \lambda_m x^{(m)}$, where $\lambda_0 \geq 0, \lambda_1 \geq 0, \ldots, \lambda_m \geq 0$, and $\lambda_0 + \lambda_1 + \ldots + \lambda_m = 1$. $\mathcal{K}$ may contain other points than those above; for example, $\mathcal{K}$ may be a disc in $\mathbb{R}^2$, while the above points belong to an inscribed triangle $P_0 P_1 P_2$.

In general we shall deal with the case in which $m = n$. Before further developing the properties of convex sets, we introduce some terms from set topology.

A point P is an *interior point* of a set $\mathcal{M}$ contained in $\mathbb{R}^n$, if there exists an n-dimensional ball, with centre at P, all of whose points lie in $\mathcal{M}$.

An *open set* is a set containing only interior points. The *interior* of a set $\mathcal{M}$, written Int $\mathcal{M}$, or $\mathcal{M}^\circ$, is the set of all its interior points.

It is easy to show that if $\mathcal{K}$ is an n-dimensional convex set in $\mathbb{R}^n$, then it must contain interior points, since the centre of gravity of an n-dimensional tetrahedron is an interior point of the tetrahedron, and we know that such a tetrahedron is contained in $\mathcal{K}$.

Theorem 1. *If $\mathcal{K}$ is an n-dimensional convex set in $\mathbb{R}^n$, then* Int $\mathcal{K}$ *is a convex set.*

Let P be any point in Int $\mathcal{K}$, $Q \in \mathcal{K}$, $Q \neq P$, and R a point in the segment PQ, $R \neq Q$. Then $R \in \mathcal{K}$, since $\mathcal{K}$ is convex. We shall prove that $R \in$ Int $\mathcal{K}$.

Let $|PQ|$ denote the length of PQ. Suppose $|PQ| = b$, and $|RQ| = a$. Since $P \in$ Int $\mathcal{K}$, there is an n-dimensional ball with radius $r > 0$ and centre P, all of whose points lie in $\mathcal{K}$. By choosing r small enough, we may further suppose that $b > r$. Construct an n-dimensional ball of radius $\frac{ra}{b}$, with R as centre, and choose a point R' in the interior of the ball. Since $b > r > 0$, we have $a > \frac{ra}{b}$,

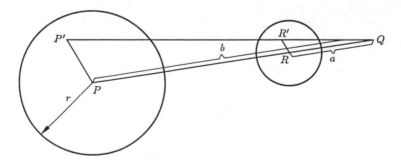

and therefore $R' \neq Q$. Construct the point P' on the ray emanating from Q and passing through R', so that $\frac{|QR'|}{|QP'|} = \frac{a}{b}$. Then P' belongs to the interior of the ball of radius r around P. Therefore P' will be a point in $\mathcal{K}$, and since Q is also in $\mathcal{K}$, we have $R' \in \mathcal{K}$. Since R' was an arbitrary point in a ball around R, we have proved that R is an interior point of $\mathcal{K}$.

§2. Convex bodies

We introduce some definitions for well-known ideas.

Definition. A *convex body* is a bounded, convex, open set in $\mathbb{R}^n$.

The interior of an n-dimensional ball, defined by

$$x_1^2 + x_2^2 + \ldots + x_n^2 < a^2 , \quad a \neq 0 ,$$

provides an example.

A *frontier point* P [cf. Lecture XIV, §1] of a convex body $\mathcal{B}$ is a point *not* belonging to $\mathcal{B}$, such that there exist points of $\mathcal{B}$ arbitrarily close to P.

The *surface* $\partial\mathcal{B}$ of a convex body $\mathcal{B}$ is the set of all its frontier points.

Let $\overline{\mathcal{B}} = \mathcal{B} \cup \partial\mathcal{B}$. Then $\overline{\mathcal{B}}$, called the *closure* of $\mathcal{B}$, is a closed set; that is, it contains all its limit points.

Theorem 2. *If $\mathcal{B}$ is a convex body, then* Int $\overline{\mathcal{B}} = \mathcal{B}$.

This theorem is not true for an arbitrary set $\mathcal{B}$. Suppose, for instance in $\mathbb{R}^2$, that $\mathcal{B}$ is the interior of the unit disc excluding the origin. Then $\overline{\mathcal{B}}$ is the closed unit disc, while Int $\overline{\mathcal{B}}$ is the complete interior of the unit disc. Of course the reason for the failure of the theorem is that the original set $\mathcal{B}$ is not convex. [If $\mathbb{Q}$ denotes the set of rational numbers in $\mathbb{R}^1$, and we define $\mathcal{U} = [0, 1] \cap \mathbb{Q}$, then Int $\overline{\mathcal{U}} = (0, 1)$, and it is not true that $\mathcal{U} \subset$ Int $\overline{\mathcal{U}}$, nor is it true that $\mathcal{U} \supset$ Int $\overline{\mathcal{U}}$.]

To prove Theorem 2, we shall first prove that $\mathcal{B} \subset$ Int $\overline{\mathcal{B}}$, and then that Int $\overline{\mathcal{B}} \subset \mathcal{B}$. Let $P \in \mathcal{B}$. Since $\mathcal{B}$ is open, there exists a ball with centre at P, which lies completely in $\mathcal{B}$. This ball belongs also to $\overline{\mathcal{B}}$, since $\overline{\mathcal{B}} \supset \mathcal{B}$. Therefore P is an interior point of $\overline{\mathcal{B}}$. Hence $\mathcal{B} \subset$ Int $\overline{\mathcal{B}}$. Conversely let $P \in$ Int $\overline{\mathcal{B}}$. Then there exists a ball, with centre at P, all of whose points lie in $\overline{\mathcal{B}}$. Inscribe in the ball an n-dimensional tetrahedron containing P in its interior. All the vertices of the tetrahedron belong to $\overline{\mathcal{B}}$, and therefore either to $\mathcal{B}$ or to $\partial\mathcal{B}$. If any vertex belongs to $\partial\mathcal{B}$, we can find points of $\mathcal{B}$ arbitrarily close to it, so that we can construct a new tetrahedron containing P in its interior, whose vertices all belong to $\mathcal{B}$. Since $\mathcal{B}$ is convex, P must belong to $\mathcal{B}$. Hence Int $\overline{\mathcal{B}} \subset \mathcal{B}$.

Theorem 3. *The closure of a convex body is convex.*

Let $\mathcal{B}$ be a convex body, with $P \in \overline{\mathcal{B}}, Q \in \overline{\mathcal{B}}$. We can then find two sequences of points $P_j \in \mathcal{B}, Q_j \in \mathcal{B}$ converging respectively to P and to Q. The segments $P_j Q_j$ lie completely in $\mathcal{B}$ and tend to the segment PQ. Therefore the points of PQ are limit points of sequences of points in $\mathcal{B}$, and so belong to $\overline{\mathcal{B}}$. This proves that $\overline{\mathcal{B}}$ is convex.

§3. Gauge function of a convex body

One of the many important ideas introduced by Minkowski into the study of convex bodies was that of gauge function. Roughly, the gauge function is the equation of a convex body. Minkowski showed that the gauge function could be defined in a purely geometric way and that it must have certain properties analogous to those possessed by the distance of a point from the origin. He also showed that conversely given any function possessing these properties, there exists a convex body with the given function as its gauge function.

Before giving the definition of gauge function, we shall investigate some further properties of the surface $\partial \mathcal{B}$ of a convex body $\mathcal{B}$. Let O be a point in $\mathcal{B}$. Consider any ray starting from O and going to infinity in an arbitrary direction. We shall prove that this ray intersects $\partial \mathcal{B}$ in exactly one point. The ray must intersect $\partial \mathcal{B}$ in at least one point, because $\mathcal{B}$ is bounded; and all points Q far enough away do not belong to $\mathcal{B}$. The distances $|OQ|$ of all points Q of the ray which do not belong to $\mathcal{B}$, have a greatest lower bound λ, say. Then the point P on the ray such that $|OP| = \lambda$ belongs to $\partial \mathcal{B}$. For if we choose any point P' between P and O, then by the construction of P, P' belongs to $\mathcal{B}$. This shows that there exist points of $\mathcal{B}$ arbitrarily close to P, but that P is not a point of $\mathcal{B}$, since there is no ball with centre at P, which is completely contained in $\mathcal{B}$. Hence $P \in \partial \mathcal{B}$.

If the ray starting from O intersects $\partial \mathcal{B}$ in (at least) two different points, first in P and then in Q, we reach a contradiction. For Theorem 3 implies that $\overline{\mathcal{B}}$ is convex since $\mathcal{B}$ is, and the proof of Theorem 1 shows that P must be an interior point of $\overline{\mathcal{B}}$ and so belong to $\mathcal{B}$. Since $\mathcal{B}$ is open, P cannot belong both to $\mathcal{B}$ and to $\partial \mathcal{B}$.

Given a convex body $\mathcal{B} \subset \mathbb{R}^n$ containing the origin O, we define a function $f : \mathbb{R}^n \to [0, \infty)$, as follows. If $x \in \partial \mathcal{B}$, (and x denotes also the vector representing the point x), then

$$(1) \qquad\qquad\qquad f(x) = 1 \ .$$

For any other vector $x \neq 0$, construct the ray through O and the point (whose vector is) x. Suppose this ray intersects the surface $\partial \mathcal{B}$ in a point y. Then there exists a $\lambda > 0$, such that $x = \lambda y$, and we define

$$(2) \qquad\qquad\qquad f(x) = \lambda \ .$$

We complete the definition of f by setting

$$(3) \qquad\qquad\qquad f(0) = 0 \ .$$

The function f so defined is the *gauge function* of the convex body $\mathcal{B}$.

We now prove that f is a positive-homogeneous (since $\lambda > 0$) function of degree one.

Theorem 4. *If f is the gauge function of a convex body $\mathcal{B} \subset \mathbb{R}^n$ containing the origin O, $x \in \mathbb{R}^n$, and $\mu > 0$, then $f(\mu x) = \mu f(x)$.*

This is trivial for $x = 0$ because of (3). If $x \neq 0$, there exists a point $y \in \partial B$, such that $x = \lambda y$, $\lambda > 0$. Because of (1) and (2), we then have

$$f(\mu x) = f(\mu \lambda y) = \mu \lambda = \mu f(x) \,.$$

We note the trivial

Theorem 5. *If f is the gauge function of a convex body $B \subset \mathbb{R}^n$ containing the origin O, $x \in \mathbb{R}^n$, then $f(x) > 0$ for $x \neq 0$, while $f(0) = 0$.*

Note that the properties of the gauge function f, as expressed in Theorems 4 and 5, are also properties of the distance function $|\,|$, which assigns to a vector $x \in \mathbb{R}^n$ (representing the point X) the distance of X from the origin, that is $|x| = |OX| = (x_1^2 + \ldots + x_n^2)^{1/2}$, where $x = (x_1, \ldots, x_n)$. The distance function is the gauge function of the n-dimensional unit ball; it has, however, a third very important property, namely it satisfies the triangle inequality. We shall show that an arbitrary gauge function also has this property.

Theorem 6. *If f is the gauge function of a convex body $B \subset \mathbb{R}^n$ containing the origin O, and $x, y \in \mathbb{R}^n$, then*

$$f(x + y) \leq f(x) + f(y) \,.$$

[This, together with the property expressed in Theorem 4, is referred to, later on, as the *convexity property* of the gauge function f.]

By Definitions (1), (2) and (3), $f(x) \leq 1$ for all $x \in \overline{B}$, and conversely, $f(x) \leq 1$ implies that $x \in \overline{B}$. Let $x', y' \in \overline{B}$. Then by Theorem 3 and the definition of a convex set, we have $\lambda x' + \mu y' \in \overline{B}$, for $\lambda > 0$, $\mu > 0$, and $\lambda + \mu = 1$, so that

$$(4) \qquad\qquad f(\lambda x' + \mu y') \leq 1 \,.$$

The theorem is trivial if either $x = 0$, or $y = 0$. Assume that $x \neq 0$, $y \neq 0$, and define

$$x^* = \frac{1}{f(x)} \cdot x \,, \qquad y^* = \frac{1}{f(y)} \cdot y \,.$$

By Theorem 4, we have $f(x^*) = f(y^*) = 1$, therefore $x^* \in \overline{B}$, $y^* \in \overline{B}$. Let

$$\lambda = \frac{f(x)}{f(x) + f(y)} \,, \qquad \mu = \frac{f(y)}{f(x) + f(y)} \,;$$

then we have, from (4), $f(\lambda x^* + \mu y^*) \leq 1$, or using Theorem 4,

$$f\left(\frac{1}{f(x) + f(y)}(x + y)\right) = \frac{f(x + y)}{f(x) + f(y)} \leq 1 \,,$$

so that we have finally $f(x + y) \leq f(x) + f(y)$.

This shows that the gauge function has all the properties of a distance function. However, the distance is measured differently in different directions. The unit distance in any direction is the distance from O to the point on the surface in that direction.

Examples

(i) Let $\mathcal{B}$ be the interior of a square in $\mathbb{R}^2$ with vertices at $(1,1)$, $(-1,1)$, $(-1,-1)$, and $(1,-1)$. Its gauge function is given by

$$f(x) = f(x_1, x_2) = \max\{|x_1|, |x_2|\} .$$

(ii) Let $\mathcal{B}$ be the elliptic disc in $\mathbb{R}^2$ with semi-major axis a, and semi-minor axis b, as shown in the figure. Its gauge function is given by

$$f(x) = f(x_1, x_2) = \left(\frac{x_1^2}{a^2} + \frac{x_2^2}{b^2} \right)^{1/2} .$$

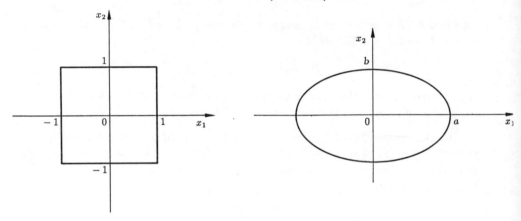

Theorem 7. *If f is any function defined on $\mathbb{R}^n$, with the properties:*

$$f(x) > 0 , \quad for\ x \in \mathbb{R}^n - \{0\} ;$$
$$f(\lambda x) = \lambda f(x) , \quad for\ \lambda > 0 , \quad x \in \mathbb{R}^n ;$$
$$f(x + y) \le f(x) + f(y) , \quad for\ x, y \in \mathbb{R}^n ,$$

then there exists a convex body $\mathcal{B}$ with f as its gauge function.

Consider the set $\{x | f(x) < 1\}$; we shall show that $\mathcal{B}$ is this set. For that purpose we must show that this set is open, convex, and bounded.

To prove that the set is open, we first show that f is continuous. Let $e^{(1)}, e^{(2)}, \ldots, e^{(n)}$ be the coordinate vectors along the coordinate axes (of a rectangular coordinate system introduced) in $\mathbb{R}^n$. Then we have

$$x = \sum_{j=1}^{n} \mu_j e^{(j)} = \sum_{j=1}^{n} \pm |\mu_j| e^{(j)} .$$

Since $|\mu_j| \geq 0$, from the third and the second properties of f,

$$f(x) \leq \sum_{j=1}^{n} f\left(\pm |\mu_j| e^{(j)} \right) = \sum_{j=1}^{n} |\mu_j| f\left(\pm e^{(j)} \right) .$$

By the second property, $f(0) = 0$. As x tends to 0, $|\mu_j| \to 0$ for $j = 1, \ldots, n$, and hence $f(x) \to 0$ as $x \to 0$, which means that f is continuous at the origin.

To prove that f is continuous at *any* point $x \in \mathbb{R}^n$, we write $x = x + y + (-y)$, and we have $f(x) \leq f(x+y) + f(-y)$, and also $f(x+y) \leq f(x) + f(y)$. Hence

$$-f(-y) \leq f(x+y) - f(x) \leq f(y) .$$

If we now let y tend to 0, then $f(y)$ and $f(-y)$ tend to zero, because f has been proved to be continuous at 0. Hence

$$f(x+y) - f(x) \to 0 , \quad \text{as } y \to 0 ,$$

which proves that f is continuous at the (arbitrary) point x, and hence on $\mathbb{R}^n$.

Since f is continuous, the set $\{x | f(x) < 1\}$ is *open*, because given any point x in the set, there exists a ball with x as centre, such that $f(y) < f(x) + \frac{1}{2}(1 - f(x)) < 1$, for any point y in the ball. [We can also say that $\{x | f(x) < 1\} = f^{-1}[0, 1)$, where $[0, 1)$ is open in $[0, \infty)$.]

To prove that the set $\{x | f(x) < 1\}$ is *convex*, suppose that x and y are in the set, so that $f(x) < 1$, $f(y) < 1$. Because of our assumptions on f, we have, for *any* positive λ, μ with $\lambda + \mu = 1$,

$$f(\lambda x + \mu y) \leq f(\lambda x) + f(\mu y) = \lambda f(x) + \mu f(y) < \lambda + \mu = 1 ,$$

so that $\lambda x + \mu y$ also belongs to the set.

If the set was not bounded, then there would exist a sequence of points x_j, such that $f(x_j) < 1$, and $|x_j| \to \infty$ as $j \to \infty$. Let $\lambda_j = \frac{1}{|x_j|}$ (for j large enough, so that $|x_j| \neq 0$), then

$$f(\lambda_j x_j) = \frac{1}{|x_j|} f(x_j) < \frac{1}{|x_j|} \to 0 , \quad \text{as } j \to \infty .$$

The point $\lambda_j x_j$ lies on the unit sphere. Since f is continuous, it attains a minimum on the unit sphere, and the minimum is strictly positive, because of the first property of f. This contradicts the fact that $f(\lambda_j x_j) \to 0$ as $j \to \infty$. Therefore the set $\{x | f(x) < 1\}$ is *bounded*, and hence defines a convex body $\mathcal{B}$. It follows that f is the gauge function of $\mathcal{B}$.

§4. Convex bodies with a centre

A point C is said to be a *centre* of a convex body $\mathcal{B}$, if whenever a point P belongs to $\mathcal{B}$, its reflection in C (that is, the point P^* defined by $P^* = P$ if $P = C$; otherwise, C is the midpoint of the segment PP^*) also belongs to $\mathcal{B}$. Obviously C is then a point of $\mathcal{B}$. There cannot be two centres O (the origin) and A (with vector a), $A \neq O$, of $\mathcal{B}$; for otherwise, by reflecting a in O, $-a$ would be in $\mathcal{B}$, then by reflection in A, $+3a$ would be in $\mathcal{B}$, and by successive further reflections in O and A, there would be an unbounded sequence of points in $\mathcal{B}$, which is impossible, since $\mathcal{B}$ is bounded.

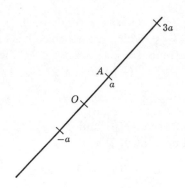

Theorem 8. *If $\mathcal{B}$ is a convex body in $\mathbb{R}^n$ with centre O, then O is also a centre of $\partial\mathcal{B}$ in the sense that if $x \in \partial\mathcal{B}$, then $-x \in \partial\mathcal{B}$.*

Let $x \in \partial\mathcal{B}$. Consider the segment joining O and x all of whose points with the exception of x belong to $\mathcal{B}$. Since O is a centre of $\mathcal{B}$, all of the points of the segment joining O and $-x$ are points of $\mathcal{B}$, except possibly the point $-x$. Now $-x$ cannot belong to $\mathcal{B}$, for otherwise $-(-x) = x \in \mathcal{B}$. This proves that the point $-x \notin \mathcal{B}$ is a limit point of a sequence of points from $\mathcal{B}$, hence belongs to $\partial\mathcal{B}$, and Theorem 8 follows.

Let f be the gauge function of a convex body $\mathcal{B}$ with centre at O, then $f(x) = f(-x)$. This is trivial for $x = 0$. For $x \neq 0$, let $x = \lambda y$, $\lambda > 0$, $y \in \partial\mathcal{B}$. Then we have $f(x) = f(\lambda y) = \lambda f(y) = \lambda$, and $f(-x) = f(-\lambda y) = \lambda f(-y) = \lambda$, since $-y \in \partial\mathcal{B}$, by Theorem 8. This proves

Theorem 9. *The gauge function of a convex body $\mathcal{B}$ with centre at O is an even function.*

The converse is also true. For after Theorem 7, an even gauge function must belong to a convex body *with* a centre at O, since $f(x) < 1$ implies that $f(-x) < 1$.

We shall now obtain the volume V of a convex body $\mathcal{B}$ with $O \in \mathcal{B}$ in terms of its gauge function.

Let x be any point of the $(n-1)$-dimensional unit sphere $S^{n-1} \subset \mathbb{R}^n$, with centre at O. The ray passing from O to x intersects ∂B in the point y, say. Then $y = \frac{1}{f(x)} \cdot x$. Let

$$r(x) = |y| = \frac{1}{f(x)} \cdot |x| = \frac{1}{f(x)} \; .$$

The function r is continuous on S^{n-1}, since f is continuous, and $f(x) \neq 0$ for $x \in S^{n-1}$. [In fact, $f(x) \neq 0$ for $x \neq 0$.] We have

$$V = \int_B dx = \int_{S^{n-1}} \int_0^{r(x)} r^{n-1} dr \, d\omega = \int_{S^{n-1}} \frac{r^n(x)}{n} d\omega = \frac{1}{n} \int_{S^{n-1}} \left(\frac{1}{f(x)} \right)^n d\omega \; ,$$

where $d\omega$ is the surface element of S^{n-1}.

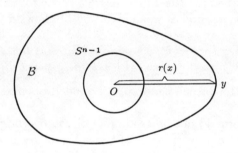

Lecture II

§1. Minkowski's First Theorem

In this section and in the later ones we shall discuss the application of the theory of convex bodies to the theory of numbers. The connexion with the theory of numbers will be found in the study of points all of whose coordinates are integers. These points will be called *integral points* or *g-points*. The set of all *g*-points will be called a *lattice*. [See also § 5.] The following theorem about lattice points and convex bodies was proved by Minkowski.

Theorem 10. *A convex body in $\mathbb{R}^n$, having a centre at the origin and having a volume larger than 2^n, must contain at least one g-point different from 0.*

The volume, in the Jordan sense, of a bounded open *set* $\mathcal{M} \subset \mathbb{R}^n$ will be defined as follows:

Consider the volume V_I of a finite number of non-overlapping n-dimensional cubes which are such that every point of the cubes belongs to $\mathcal{M}$ (for example, the volume of the cubes bounded by A, B, C, D, E, F in the diagram). Compare it with V_E, the volume of any finite set of non-overlapping cubes which cover $\mathcal{M}$, that is, every point of $\mathcal{M}$ belongs to one of the cubes, and every cube contains at least one point of $\mathcal{M}$ (for example, the cubes bounded by $A', B', C', D', E', F', G', H'$).

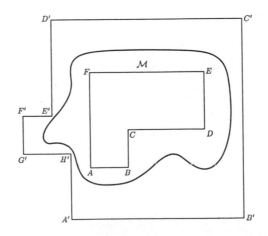

If as the maximum length of the sides of these cubes approaches zero, the difference $V_E - V_I$ can be made as small as we please, then the *volume* of $\mathcal{M}$ exists, and is the common limit of V_E and V_I. [If the volume of $\mathcal{M}$ exists, then the characteristic function of $\mathcal{M}$ is Riemann integrable.]

§2. Lemma on bounded open sets in $\mathbb{R}^n$

Before discussing Minkowski's theorem, we shall prove a lemma, which is important in itself, and which will give us a quick proof of Theorem 10.

Lemma 1. *Let $\mathcal{M}$ be a bounded open set in $\mathbb{R}^n$, with volume greater than 1. Then $\mathcal{M}$ contains two distinct points x and y, such that*

$$x_i - y_i = an\ integer \quad (for\ i = 1, \ldots, n) \ ,$$

where x_i and y_i are the coordinates of the points x and y respectively.

Note that $x - y$ is a g-point.

Geometrically, Lemma 1 states that for any bounded open set $\mathcal{M}$ with volume greater than 1, we can translate the lattice of g-points, so that $\mathcal{M}$ will contain at least two points of the lattice, for example P and Q in diagram (i).

If we translate the lattice back to the original position, meanwhile leaving the set $\mathcal{M}$ unchanged, then P and Q become points of the unit cubes whose vertices are the g-points, and P and Q will be similarly placed relative to the sides of the cubes to which they belong [see diagram (ii)].

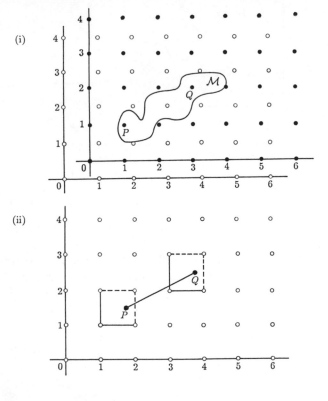

Let $\mathcal{E}$ be the cube defined by $0 \le x_k < 1$, $k = 1, \ldots, n$. With $\mathcal{E}$ is associated the g-point $(0, \ldots, 0)$. We adopt the same convention for all cubes, so that each cube is associated to one particular vertex.

If the cubes are made to coincide by a translation, the points P and Q will also be coincident [see diagram (ii)].

These considerations will serve as a basis for the proofs of Lemma 1. We give three proofs.

First proof. Let φ be the characteristic function of $\mathcal{M}$, that is to say, $\varphi(x) = 1$ if $x \in \mathcal{M}$, and $\varphi(x) = 0$ if $x \notin \mathcal{M}$. Note that since $\mathcal{M}$ has a volume, φ is integrable. Consider the function

$$\psi(x) = \sum_g \varphi(x + g) \,,$$

where the sum is over all g-points. If x belongs to the cube $\mathcal{E}$, then $\varphi(x+g) = 0$, for sufficiently large values of $|g|$, since $\mathcal{M}$ is bounded. This shows that $\psi(x)$, for x in $\mathcal{E}$, is the sum of only a finite number of non-zero terms, each such term being the characteristic function for the intersection of the set $\mathcal{M}$ with the unit cube whose vertex is g.

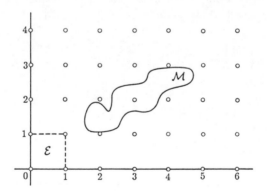

In the diagram the non-zero terms will be given by $g = (1,1), (2,1), (3,1), (2,2)$, $(3,2)$, and $(4,2)$.

From the definition of volume it follows that these intersections will have a volume, and so their characteristic function will be integrable. Now remembering that $\psi(x)$ contains only a finite number of non-zero terms, so that the interchange of integration and summation is justified, we have

$$I = \int_{\mathcal{E}} \psi(x)\, dx = \sum_g \int_{\mathcal{E}} \varphi(x+g)\, dx = \sum_g \int_{\mathcal{E}+g} \varphi(y)\, dy \,,$$

where $y = x + g$, and $\mathcal{E} + g$ is the cube $\mathcal{E}$ translated by the vector g. As g runs through all g-points, $\{\mathcal{E} + g\}$ completely fills the whole of space, so that

$$I = \int_{\mathbb{R}^n} \varphi(y)\, dy = \int_{\mathcal{M}} dy = V \,,$$

where V is the volume of $\mathcal{M}$.

From its definition, $\psi(x)$ is a non-negative, bounded integer for $x \in \mathcal{E}$; therefore it must attain a maximum m, say, in $\mathcal{E}$, but then

$$I = \int_{\mathcal{E}} \psi(x)\, dx \le m \cdot 1 \,,$$

so that

$$m \ge I = V > 1 \,,$$

by our hypothesis on $\mathcal{M}$. Since m is an integer, it must be at least two. This means there exists a point x' in $\mathcal{E}$, such that $\psi(x') \ge 2$, which implies that there exist two distinct g-points, say g' and g'', such that $x' + g'$ and $x' + g''$ belong to $\mathcal{M}$. This proves the lemma, since we can take $x = x' + g'$, $y = x' + g''$.

Second proof. This proof ist just a geometric interpretation of the first proof. Let $\mathcal{M}_g$ represent the intersection of $\mathcal{M}$ and the cube $\mathcal{E} + g$, and V_g the volume of $\mathcal{M}_g$. We have $\mathcal{M} = \bigcup_g \mathcal{M}_g$, and $V = \sum_g V_g$. Note that since $\mathcal{M}$ is bounded, only a finite number of the $\mathcal{M}_g$ are non-empty, and the above sum contains only a finite number of non-zero terms.

Translate the set $\mathcal{M}$ by the vector $-g$. The intersection of $\mathcal{E}$ and the set $\mathcal{M} - g$ is the set $\mathcal{M}_g - g$ with volume V_g. All such sets lie in $\mathcal{E}$. Since $V = \sum_g V_g > 1$, and $\mathcal{E}$ has volume 1, at least two of the sets $\mathcal{M}_g - g$ overlap. Suppose x' is a common point of the sets $\mathcal{M}_{g'} - g'$ and $\mathcal{M}_{g''} - g''$, with $g' \ne g''$. Then the points $x' + g'$ and $x' + g''$ belong to $\mathcal{M}$, and their difference is a non-zero g-point.

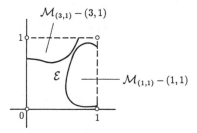

(Compare with the last diagram)

Third proof. In this proof we prove a little more than what the lemma states; we show that we can find x and y such that they will have rational coordinates.

Let $\tau = 1, 2, \ldots$. Consider the cubes $(\mathcal{E} + g)/\tau$, that is, the unit cube $\mathcal{E}$, translated by the vector g, and then contracted in the ratio $1 : \tau$. Suppose that the number of these cubes which lie completely in $\mathcal{M}$ is $A(\tau)$. Since $A(\tau) \cdot \tau^{-n}$ is the total volume of these cubes, we have, from the definition of the volume V of $\mathcal{M}$,

$$\lim_{\tau \to \infty} A(\tau) \cdot \tau^{-n} = V > 1 .$$

This implies that there exists a number τ', such that $A(\tau') > (\tau')^n$; that is, the number of cubes with sides of length $\frac{1}{\tau'}$ which we consider is greater than $(\tau')^n$.

To the origin in the cube $\mathcal{E}$ corresponds the point g in $\mathcal{E} + g$, and the point g/τ' in $(\mathcal{E} + g)/\tau'$. We have just shown that there are more than $(\tau')^n$ values of g for which g/τ' is a point of $\mathcal{M}$. Let $g_i, i = 1, \ldots, n$, be the coordinates of any one of these g's. Consider the least positive residue of g_i modulo τ'. [The least positive residue of an integer a modulo the integer $b (> 0)$ is the integer r such that $a - r$ is divisible by b and $0 \leq r < b$. There are exactly b residue classes, defined by the values $r = 0, 1, \ldots, b-1$.] For each g_i there are τ' possible residue classes, and for the vector g, there are only $(\tau')^n$ possibilities. Therefore, from the definition of τ', there must exist two different points, g' and g'', "belonging to the same residue class modulo τ'", that is $g_i' - g_i''$ is divisible by τ' for $i = 1, \ldots, n$. Since g'/τ' and g''/τ' belong to $\mathcal{M}$, our lemma is proved if we take $x = g'/\tau'$ and $y = g''/\tau'$, since $x - y = (g' - g'')/\tau'$, a g-point.

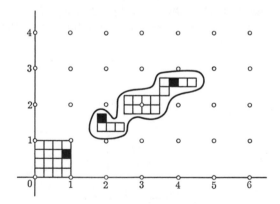

§3. Proof of Minkowski's First Theorem

The proof of Theorem 10 will be reduced to an application of the lemma of § 2. Let B be a convex body with O as centre, and with a volume $V > 2^n$. Then we shall prove that there exists in B a g-point different from the origin.

Contract B by the factor $\frac{1}{2}$, that is, multiply the coordinates of every point of B by $\frac{1}{2}$. We shall have a new convex body with volume $V' > 1$. Call this body M. By the lemma proved in §2, there exist two points x and y in M, such that $x - y = g \neq 0$, where g is a g-point. Then $2x$ and $2y$ belong to B. Since B has O as centre, $-2y$ belongs to B, and since B is convex, the midpoint of $2x$ and $-2y$ belongs to B. That is, $\frac{1}{2}(2x) + \frac{1}{2}(-2y) = x - y = g$ belongs to B, which proves the theorem.

It seems that full use has not been made of the convexity of B. This is not so, for it can be proved that if M is a bounded open set, and if whenever P and $Q \neq P$ belong to M, the midpoint of the segment PQ belongs to M, then the entire segment PQ belongs to M, hence M is a convex body.

Suppose V is exactly equal to 2^n. Then there exists a non-zero g-point either in B or in ∂B. This can be shown as follows.

Let λ be a real number between 1 and 2. Stretch B in the ratio $\lambda : 1$, and denote the convex body so formed by λB. Since the volume of λB is $2^n \lambda^n > 2^n$, an application of Minkowski's theorem shows that there exists an integral point $g_\lambda \neq 0$ in λB. Since λB is contained in $2B$, g_λ is always contained in $2B$, and therefore there are only a finite number of possibilities for g_λ. Let λ approach 1 in such a way that the sequence g_λ is always the same point $g \neq 0$. Now $\frac{1}{\lambda}g$ belongs to B, and as $\lambda \to 1$, $\frac{1}{\lambda}g \to g$, so that g is a limit point of B. Therefore g belongs either to B or to ∂B.

The constant in the theorem cannot be smaller than 2^n, for consider the cube, the coordinates of whose points lie in $(-\delta, \delta)$, where $0 < \delta < 1$, so that the volume of the cube is $(2\delta)^n$. If x_k, the kth coordinate of any point of the cube is an integer, then $x_k = 0$. This shows that the only point with integral coordinates inside this cube is O. Since the volume can be made arbitrarily close to 2^n for δ near enough to 1, this shows that the volume cannot be taken smaller than 2^n.

§ 4. Minkowski's theorem for the gauge function

We can now translate Theorem 10 into a theorem about the values taken by the gauge function.

Theorem 11. *Given an even gauge function $f : \mathbb{R}^n \to [0, \infty)$, let V be the volume of the convex body $B = \{x | f(x) < 1\}$. If $V \geq 2^n$, then there exists a g-point, say g, $g \neq 0$, such that $f(g) \leq 1$.*

By the converse of Theorem 9, O must be the centre of B; and Theorem 11 follows from Theorem 10 together with the remark, in italics, of the previous section.

Theorem 12. *Given an even gauge function $f : \mathbb{R}^n \to [0, \infty)$, let μ be the minimum of the values taken by $f(x)$, when x goes through all g-points different from the origin. Let V denote the volume of the convex body $\mathcal{B} = \{x | f(x) < 1\}$. Then we have*

$$\mu^n V \leq 2^n .$$

For $\lambda > 0$, let $\mathcal{B}_\lambda$ denote the convex body $\{x | f(x) < \lambda\}$. The volume of $\mathcal{B}_\lambda$ is $\lambda^n V$, and if $0 < \lambda_1 < \lambda_2$, then $\mathcal{B}_{\lambda_1} \subset \mathcal{B}_{\lambda_2}$. [Since there are only a finite number of g-points contained in any $\mathcal{B}_\lambda$, we conclude that f attains a minimum on the set of g-points different from 0.]

Let λ_0 denote the least upper bound of all those λ for which $\mathcal{B}_\lambda$ contains no g-point different from 0. Since the convex body $\mathcal{B}_{\lambda_0}$ is open, it contains no g-point different from 0, and by Theorem 10, we have

$$\lambda_0^n V \leq 2^n .$$

To prove the theorem, it is sufficient to show that $\mu \leq \lambda_0$.

If $\mu > \lambda_0$, we can find an $\epsilon > 0$, such that $\mu > \lambda_0 + \epsilon$. And $\mathcal{B}_{\lambda_0 + \epsilon}$ contains a g-point, say g, different from 0, because of the definition of λ_0. And $f(g) < \lambda_0 + \epsilon < \mu$, which contradicts the definition of μ. Hence $\mu \leq \lambda_0$. [If $\mu < \lambda_0$, then there exists a g-point $g \neq 0$, such that $f(g) = \mu$, hence $g \in \mathcal{B}_{\lambda_0}$, which is impossible, by definition of λ_0. It follows that $\mu = \lambda_0$.]

§5. The minimum of the gauge function for an arbitrary lattice in $\mathbb{R}^n$

Instead of the coordinates $x_1, x_2, \ldots, x_n$ we introduce in $\mathbb{R}^n$ new coordinates defined by the equations

$$y_1 = a_{11}x_1 + a_{12}x_2 + \ldots + a_{1n}x_n + c_1 ,$$
$$y_2 = a_{21}x_1 + a_{22}x_2 + \ldots + a_{2n}x_n + c_2 ,$$
$$\vdots$$
$$y_n = a_{n1}x_1 + a_{n2}x_2 + \ldots + a_{nn}x_n + c_n ,$$

where we *assume* that the determinant of the transformation $x \mapsto y$ is *not* zero. Let $D = |\det(a_{jk})|$.

Since $D > 0$, we can solve for the x_j in terms of the y_k, for $j, k = 1, \ldots, n$. Therefore the transformation sets up a one-to-one correspondence between the x-space and the y-space. Segments go over into segments, open sets into open sets, and convex sets into convex sets.

From now on we suppose that $c_j = 0$, $j = 1, \ldots, n$.

The image of the set of all g-points in the x-space is (also) called a *lattice* $\wedge$ in the y-space, and we refer to $|\det(a_{jk})|$ as the determinant of the lattice.

If 0 is the centre of a convex body $\mathcal{A}$ in the x-space [looked upon now not just as an affine space, but as a vector space], then 0 is also the centre of the convex body $\mathcal{B}$ in the y-space, which is the image of $\mathcal{A}$ under the above mapping. This follows from the fact that if y corresponds to x, then $-y$ corresponds to $-x$, so that if $-x \in \mathcal{A}$ when $x \in \mathcal{A}$, then $-y \in \mathcal{B}$ when $y \in \mathcal{B}$.

Let $y \mapsto f(y)$ be the gauge function for $\mathcal{B}$. If we substitute for y its value in terms of x, $y \mapsto f(y)$ becomes some function $x \mapsto f_1(x)$, and $f_1(x)$ is the gauge function for $\mathcal{A}$, because $f(y) < 1$ for $y \in \mathcal{B}$ implies that $f_1(x) < 1$ for $x \in \mathcal{A}$ and vice versa.

Let V denote the volume of $\mathcal{B}$, and U that of $\mathcal{A}$. Then since D is the Jacobian of the transformation of the x-space into the y-space, it follows that

$$V = \int_{\mathcal{B}} dy = \int_{\mathcal{A}} D\, dx = DU \ .$$

Now, by Theorem 12, if μ is the minimum of $f_1(x)$, when x runs through the g-points different from the origin, then

$$\mu \le 2U^{-1/n} = 2(D/V)^{1/n} \ .$$

Note that because of the definition of f_1, μ is also the minimum of $f(y)$ when y runs through the set of points defined by

$$y_j = \sum_{k=1}^{n} a_{jk} g_k \ , \quad j = 1, \ldots, n \ ,$$

where the g_k are integers not all of which are zero. Hence we have proved

Theorem 13. *Given an even gauge function $f : \mathbb{R}^n \to [0, \infty)$, and a non-singular $n \times n$ matrix (a_{jk}), let $\mu = \min f(y)$, when y runs through the set of points defined by*

$$y = (y_1, \ldots, y_n) \ , \quad y_j = \sum_{k=1}^{n} a_{jk} g_k \ , \quad j = 1, \ldots, n \ ,$$

where the g_k are integers not all of which are zero. Then we have

$$\mu^n V \le 2^n D \ ,$$

where V is the volume of the convex body defined by $\{y | f(y) < 1\}$ and $D = |\det(a_{jk})|$.

Geometrically the theorem gives an upper bound for the minimum of $f(y)$, when y runs through all non-zero points of an arbitrary lattice defined by the vectors $(a_{11}, a_{21}, \ldots, a_{n1}), (a_{12}, a_{22}, \ldots, a_{n2}), \ldots, (a_{1n}, a_{2n}, \ldots, a_{nn})$. D is the volume of the unit cell of the lattice. In the special case when the lattice reduces to that of the g-points, $D = 1$ and Theorem 13 reduces to Theorem 12.

As an illustration of a general lattice, consider in $\mathbb{R}^2$ the lattice generated by the vectors $(5,3)$ and $(-1,4)$.

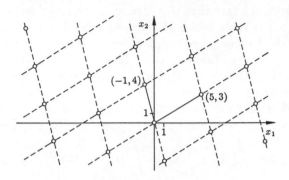

§6. Examples

We shall apply the preceding theory to the study of some special functions. Let

$$f(x) = \left(\sum_{j=1}^{n} |x_j|^r \right)^{1/r} \quad , \quad r \geq 1 , \quad x \in \mathbb{R}^n .$$

We shall prove that f is an even gauge function by showing that f possesses the three properties described in Theorem 7.

The property that $f(x) \geq 0$ is trivial, since all the terms (in the defining sum) are non-negative.

The second property, that $f(\lambda x) = \lambda f(x)$, $\lambda > 0$, is also trivial, since $f(x)$ is a positive-homogeneous function of degree one.

The third property that $f(x + y) \leq f(x) + f(y)$ is definitely not trivial. It is called *Minkowski's inequality*. We shall derive it from the inequality of the arithmetic and geometric means, and, on the way, prove *Hölder's inequality*.

Inequality of the arithmetic and geometric means. Let $p_j \geq 0$, for $j = 1, \ldots, n$, and $\sum_{j=1}^{n} p_j = 1$. For $x_j > 0$, $j = 1, \ldots, n$, let

(1)
$$M_r = \left(\sum_{j=1}^{n} p_j x_j^r \right)^{1/r} \quad , \quad r > 0 ,$$

and

$$M_0 = \prod_{j=1}^{n} x_j^{p_j} .$$

We shall prove that

(2) $$M_r \le M_{2r} \, ,$$

and also that

(3) $$\lim_{r \downarrow 0} M_r = M_0 \, .$$

We shall see that (2) is a consequence of *Schwarz's inequality*, namely: if a_j, b_j are real numbers, for $j = 1, \ldots, n$, then

(4) $$\left(\sum_{j=1}^{n} a_j b_j \right)^2 \le \left(\sum_{j=1}^{n} a_j^2 \right) \cdot \left(\sum_{j=1}^{n} b_j^2 \right) \, .$$

This follows from the algebraic identity

$$\left(\sum_{j=1}^{n} a_j^2 \right) \cdot \left(\sum_{j=1}^{n} b_j^2 \right) - \left(\sum_{j=1}^{n} a_j b_j \right)^2 = \frac{1}{2} \sum_{j=1}^{n} \sum_{k=1}^{n} (a_j b_k - a_k b_j)^2 \ge 0 \, .$$

If we set

$$a_j = \sqrt{p_j} \, , \quad b_j = \sqrt{p_j} \cdot x_j^r \, , \quad \text{for } j = 1, \ldots, n \, ,$$

then we have

(5) $$\left(\sum_{j=1}^{n} p_j x_j^r \right)^2 \le \left(\sum_{j=1}^{n} p_j \right) \cdot \left(\sum_{j=1}^{n} p_j x_j^{2r} \right) = \sum_{j=1}^{n} p_j x_j^{2r} \, ,$$

since $\sum_{j=1}^{n} p_j = 1$ by assumption. Writing it differently, we get

$$M_r^{2r} \le M_{2r}^{2r} \, ,$$

so that $M_r \le M_{2r}$, which proves inequality (2).

Noting that $x_j > 0$ for $j = 1, \ldots, n$, we have

(6) $$\log M_r = \frac{1}{r} \log \left(\sum_{j=1}^{n} p_j x_j^r \right) = \frac{1}{r} \log \left(1 + \sum_{j=1}^{n} p_j (x_j^r - 1) \right) \, ,$$

since $\sum_{j=1}^{n} p_j = 1$. We have

$$x_j^r - 1 = e^{r \log x_j} - 1 = r \log x_j + O(r^2) \, , \quad (r \downarrow 0) \, ,$$

and therefore

$$\log M_r = \frac{1}{r} \log \left(1 + r \sum_{j=1}^{n} p_j \log x_j + O(r^2) \right) \, , \quad (r \downarrow 0) \, .$$

By using Taylor's formula for the logarithm on the right-hand side, we see that

$$(7) \qquad \log M_r = \sum_{j=1}^{n} p_j \log x_j + O(r) , \quad (r \downarrow 0) ,$$

hence

$$\lim_{r \downarrow 0}(\log M_r) = \sum_{j=1}^{n} p_j \log x_j = \log \left(\prod_{j=1}^{n} x_j^{p_j} \right) = \log M_0 ,$$

and because of the continuity of the logarithmic function,

$$\lim_{r \downarrow 0} M_r = M_0 ,$$

which proves (3).

For $r = 1$, M_1 is the weighted arithmetic mean $\sum_{j=1}^{n} p_j x_j$; and for $r = 0$, M_0 is the weighted geometric mean $\prod_{j=1}^{n} x_j^{p_j}$. We have proved that

$$\sum_{j=1}^{n} p_j x_j = M_1 \geq M_{\frac{1}{2}} \geq M_{\frac{1}{4}} \geq \ldots \geq M_0 = \prod_{j=1}^{n} x_j^{p_j} .$$

The inequality $\sum_{j=1}^{n} p_j x_j \geq \prod_{j=1}^{n} x_j^{p_j}$ holds trivially if any x_j, $j = 1, \ldots, n$ is zero. *Hence if $x_j \geq 0$, $p_j \geq 0$, for $j = 1, \ldots, n$, and $\sum_{j=1}^{n} p_j = 1$, then we have*

$$(8) \qquad \sum_{j=1}^{n} p_j x_j \geq \prod_{j=1}^{n} x_j^{p_j} .$$

Hölder's inequality. If $n = 2$, $p_1 = p$, $0 < p < 1$, and $p_2 = 1 - p$, in (8), then we have

$$(9) \qquad px + (1 - p)y \geq x^p \cdot y^{1-p} ,$$

for $x \geq 0, y \geq 0$.

For $j = 1, \ldots, n$, let $x_j \geq 0$, $y_j \geq 0$. If we apply inequality (9) to every pair x_j, y_j, and sum over j, we get

$$p \sum_{j=1}^{n} x_j + (1 - p) \sum_{j=1}^{n} y_j \geq \sum_{j=1}^{n} x_j^p \cdot y_j^{1-p} .$$

If, in addition,

$$(10) \qquad \sum_{j=1}^{n} x_j = 1 , \quad \sum_{j=1}^{n} y_j = 1 ,$$

we get

$$(11) \qquad \sum_{j=1}^{n} x_j^p \cdot y_j^{1-p} \leq 1 .$$

If $\sum_{j=1}^n x_j \neq 0$, $\sum_{j=1}^n y_j \neq 0$, then by replacing x_j by $x_j/\sum_{j=1}^n x_j$ and y_j by $y_j/\sum_{j=1}^n y_j$, we obtain from (11) the inequality

$$(12) \qquad \sum_{j=1}^n x_j^p \cdot y_j^{1-p} \leq \left(\sum_{j=1}^n x_j \right)^p \cdot \left(\sum_{j=1}^n y_j \right)^{1-p} .$$

If $\sum_{j=1}^n x_j = 0$, however, we must have $x_j = 0$ for $j = 1, \ldots, n$, since we have assumed that $x_j \geq 0$, in which case (12) is trivial. The same holds for y_j, so that we can formulate the result as follows: if $0 < p < 1$, and $x_j \geq 0$, $y_j \geq 0$, for $j = 1, \ldots, n$, then

$$(13) \qquad \sum_{j=1}^n x_j^p \cdot y_j^{1-p} \leq \left(\sum_{j=1}^n x_j \right)^p \cdot \left(\sum_{j=1}^n y_j \right)^{1-p} .$$

Minkowski's inequality. Let $x_j \geq 0$, $y_j \geq 0$, for $j = 1, \ldots, n$, and set $x_j + y_j = z_j$. We write

$$(14) \qquad \sum_{j=1}^n z_j^r = \sum_{j=1}^n z_j \cdot z_j^{r-1} = \sum_{j=1}^n x_j \cdot z_j^{r-1} + \sum_{j=1}^n y_j \cdot z_j^{r-1} ,$$

and apply inequality (13) to the last two sums. For $r > 1$, $p = \frac{1}{r}$, $1 - p = \frac{r-1}{r}$, we get

$$\sum_{j=1}^n x_j \cdot z_j^{r-1} = \sum_{j=1}^n (x_j^r)^{\frac{1}{r}} \cdot (z_j^r)^{\frac{r-1}{r}} \leq \left(\sum_{j=1}^n x_j^r \right)^{\frac{1}{r}} \cdot \left(\sum_{j=1}^n z_j^r \right)^{\frac{r-1}{r}} ,$$

and a similar inequality for $\sum_{j=1}^n y_j \cdot z_j^{r-1}$, hence

$$\sum_{j=1}^n z_j^r \leq \left\{ \left(\sum_{j=1}^n x_j^r \right)^{\frac{1}{r}} + \left(\sum_{j=1}^n y_j^r \right)^{\frac{1}{r}} \right\} \cdot \left(\sum_{j=1}^n z_j^r \right)^{\frac{r-1}{r}} , \qquad r > 1 ,$$

which simplifies to

$$(15) \qquad \left(\sum_{j=1}^n z_j^r \right)^{1/r} \leq \left(\sum_{j=1}^n x_j^r \right)^{1/r} + \left(\sum_{j=1}^n y_j^r \right)^{1/r} , \qquad r > 1 .$$

Substituting $z_j = x_j + y_j$ on the left-hand side, we get Minkowski's inequality, namely: if $x_j \geq 0$, $y_j \geq 0$, for $j = 1, \ldots, n$, and $r > 1$, then

$$(16) \qquad \left\{ \sum_{j=1}^n (x_j + y_j)^r \right\}^{1/r} \leq \left(\sum_{j=1}^n x_j^r \right)^{1/r} + \left(\sum_{j=1}^n y_j^r \right)^{1/r} .$$

We can now prove that the function

$$f_r : x = (x_1, \ldots, x_n) \mapsto f_r(x) = \left(\sum_{j=1}^n |x_j|^r \right)^{1/r} \quad , \quad r \geq 1 \; ,$$

is an even gauge function. We have to verify the following properties:

(i) $f_r(x) > 0$, for $x \in \mathbb{R}^n - \{0\}$;

(ii) $f_r(\lambda x) = \lambda f_r(x)$, for $\lambda > 0$, $x \in \mathbb{R}^n$;

(iii) $f_r(x + y) \leq f_r(x) + f_r(y)$, for $x, y \in \mathbb{R}^n$; and

(iv) $f_r(-x) = f_r(x)$, for $x \in \mathbb{R}^n$.

Properties (i), (ii) and (iv) are obvious. To verify (iii), we have only to note that $|x_j + y_j| \leq |x_j| + |y_j|$, and that for $r > 1$,

(17)
$$f_r(x + y) = \left(\sum_{j=1}^n |x_j + y_j|^r \right)^{1/r} \leq \left\{ \sum_{j=1}^n (|x_j| + |y_j|)^r \right\}^{1/r}$$

$$\leq \left(\sum_{j=1}^n |x_j|^r \right)^{1/r} + \left(\sum_{j=1}^n |y_j|^r \right)^{1/r} = f_r(x) + f_r(y) \; ,$$

because of (16).

If $r = 1$, then (17) is a trivial extension of the triangle inequality, and the convex body $\{x | f_1(x) < 1\}$ is the n-dimensional unit octahedron, i.e. its vertices lie on the unit sphere $S^{n-1} \subset \mathbb{R}^n$. It can be shown that its volume is $(2^n/n!)$. We shall prove this and more in Lecture III.

Applying Theorem 13 to the n-dimensional octahedron, we obtain

Theorem 14. *Given n linear forms*

$$y_1 = a_{11}x_1 + a_{12}x_2 + \ldots + a_{1n}x_n$$
$$y_2 = a_{21}x_1 + a_{22}x_2 + \ldots + a_{2n}x_n$$
$$\vdots$$
$$y_n = a_{n1}x_1 + a_{n2}x_2 + \ldots + a_{nn}x_n$$

with a non-singular matrix (a_{jk}), there exist integral values of $x_1, \ldots, x_n$, not all zero, such that

$$|y_1| + |y_2| + \ldots + |y_n| \leq (n!D)^{1/n} \; ,$$

where $D = |\det(a_{jk})|$.

For $n = 2$, the theorem states the following: given real numbers a, b, c, d, such that $D = ad - bc > 0$, there always exist integers x_1, x_2, not both zero, such that

$$|ax_1 + bx_2| + |cx_1 + dx_2| \leq \sqrt{2D} \; .$$

To see that this is not trivial, the reader may consider the case in which $a = \sqrt{7}$, $b = \sqrt{6}$, $c = \sqrt{15}$, $d = \sqrt{13}$.

Lecture III

§1. Evaluation of a volume integral

We proved in Lecture II that if f_r is defined by

$$(1) \qquad f_r(x) = \left(\sum_{j=1}^{n} |x_j|^r \right)^{1/r} , \qquad x \in \mathbb{R}^n , \quad r \geq 1 ,$$

then it is an even gauge function on $\mathbb{R}^n$. In this section we shall evaluate the integral for the volume V_r of the convex body $\mathcal{B}_r$ defined by $\{x | f_r(x) < 1\}$. We have

$$(2) \qquad V_r = \int_{\mathcal{B}_r} dx = \int \cdots \int_{\sum_{j=1}^{n} |x_j|^r < 1} dx_1 \ldots dx_n .$$

If we define

$$(3) \qquad W_{n,r} = \int \cdots \int_{\substack{\sum_{j=1}^{n} x_j^r < 1 \\ x_j \geq 0, \; j=1,\ldots,n}} dx_1 \ldots dx_n ,$$

then since $\mathcal{B}_r$ has 0 as centre, we have

$$(4) \qquad V_r = 2^n \cdot W_{n,r} .$$

For $\lambda > 0$, we have

$$\int \cdots \int_{\substack{\sum_{j=1}^{n} x_j^r < \lambda \\ x_j \geq 0, \; j=1,\ldots,n}} dx_1 \ldots dx_n = \int \cdots \int_{\substack{\sum_{j=1}^{n} \left(x_j \lambda^{-1/r} \right)^r < 1 \\ x_j \geq 0, \; j=1,\ldots,n}} dx_1 \ldots dx_n = \lambda^{n/r} \cdot W_{n,r} ,$$

as can be seen from the substitution $x_j \lambda^{-1/r} = y_j, \; j = 1, \ldots, n$. Hence

$$W_{n,r} = \int \cdots \int\limits_{\substack{x_1^r + \ldots + x_n^r < 1 \\ x_j \geq 0,\ j=1,\ldots,n}} dx_1 \ldots dx_n = \int_0^1 \left\{ \int \cdots \int\limits_{\substack{x_1^r + \ldots + x_{n-1}^r < 1 - x_n^r \\ x_j \geq 0,\ j=1,\ldots,n-1}} dx_1 \ldots dx_{n-1} \right\} dx_n$$

(5)

$$= \int_0^1 W_{n-1,r} \cdot (1 - x_n^r)^{(n-1)/r} dx_n .$$

Since $W_{n-1,r}$ is a constant with respect to x_n, it is only necessary to evaluate the integral

$$\int_0^1 (1 - x_n^r)^{(n-1)/r} dx_n .$$

Let $x_n = t^{1/r}$, so that $dx_n = \frac{1}{r} \cdot t^{(1-r)/r} dt$, and

$$\int_0^1 (1 - x_n^r)^{(n-1)/r} dx_n = \int_0^1 (1 - t)^{(n-1)/r} \cdot \frac{1}{r} \cdot t^{(1-r)/r} dt$$

$$\overset{*)}{=} \frac{\Gamma\left(\frac{n-1}{r} + 1\right) \frac{1}{r}\Gamma\left(\frac{1}{r}\right)}{\Gamma\left(\frac{n}{r} + 1\right)} = \frac{\Gamma\left(\frac{n-1}{r} + 1\right) \Gamma\left(\frac{1}{r} + 1\right)}{\Gamma\left(\frac{n}{r} + 1\right)} .$$

On using this in (5), we get

(6) $$W_{n,r} = W_{n-1,r} \cdot \frac{\Gamma\left(\frac{n-1}{r} + 1\right) \Gamma\left(\frac{1}{r} + 1\right)}{\Gamma\left(\frac{n}{r} + 1\right)} .$$

This is valid with n replaced successively by $n-1, n-2, \ldots$, and 2. Substituting the value of $W_{n-1,r}$ in terms of $W_{n-2,r}$, and then the value of $W_{n-2,r}$ in terms of $W_{n-3,r}$, and so on, we find that

(7) $$W_{n,r} = W_{1,r} \cdot \frac{\left\{\Gamma\left(\frac{1}{r} + 1\right)\right\}^n}{\Gamma\left(\frac{n}{r} + 1\right)} = \frac{\left\{\Gamma\left(\frac{1}{r} + 1\right)\right\}^n}{\Gamma\left(\frac{n}{r} + 1\right)} ,$$

hence, from (4),

(8) $$V_r = \frac{2^n \cdot \left\{\Gamma\left(\frac{1}{r} + 1\right)\right\}^n}{\Gamma\left(\frac{n}{r} + 1\right)} .$$

In particular, we have

(9) $$V_1 = \frac{2^n}{n!} , \quad \text{and} \ \ V_2 = \frac{\pi^{n/2}}{\Gamma\left(\frac{n}{2} + 1\right)} = \frac{2 \cdot \pi^{n/2}}{n\Gamma\left(\frac{n}{2}\right)} .$$

(V_1 was mentioned in Lecture II) $\left(\Gamma\left(\frac{1}{2}\right) = \sqrt{\pi}\right)$

---

*) This is an illustration of the Beta integral, which is defined as follows: for $p > 0, q > 0$,
$B(p,q) = \int_0^1 t^{p-1}(1 - t)^{q-1} dt = \frac{\Gamma(p)\Gamma(q)}{\Gamma(p+q)}$, whence $\Gamma(p+1) = p\Gamma(p)$.

§2. Discriminant of an irreducible polynomial

The results of Lecture II about the "minimum" of a gauge function will be used to obtain a lower bound for the discriminant of a polynomial which is irreducible over $\mathbb{Q}$, the field of rational numbers. Before stating the theorem, we recall some facts from algebra [for details see Bôcher's *Introduction to Higher Algebra*].

A polynomial $P(\xi) = \xi^n + a_1\xi^{n-1} + \ldots + a_n$, with rational coefficients a_j, $j = 1, \ldots, n$, is said to be irreducible (over $\mathbb{Q}$), if it cannot be expressed as the product of two non-constant polynomials with rational coefficients. This implies that any zero of P is not a zero of any polynomial of lower degree, not identically zero, with rational coefficients. Also, every zero of P is simple, for otherwise it would be a zero of P', the derivative of P, which is of lower degree, and does not vanish identically.

If $\xi_1, \ldots, \xi_n$ denote the zeros of P, then the *discriminant* Δ of P is defined by

$$\Delta = \prod_{1 \leq j < k \leq n} (\xi_j - \xi_k)^2 .$$

It can be shown that Δ is a polynomial of degree $2n - 2$ in the coefficients $a_1, \ldots, a_n$, and that

$$(10) \qquad \Delta = \det \begin{pmatrix} \xi_1^{n-1} & \xi_1^{n-2} & \cdots & 1 \\ \xi_2^{n-1} & \xi_2^{n-2} & \cdots & 1 \\ \vdots & \vdots & & \vdots \\ \xi_n^{n-1} & \xi_n^{n-2} & \cdots & 1 \end{pmatrix}^2 .$$

Furthermore, let $Q(x_1, \ldots, x_n)$ be a polynomial with integral coefficients which is symmetric in $x_1, \ldots, x_n$. Then $Q(\xi_1, \ldots, \xi_n)$ can be expressed as a polynomial in $a_1, \ldots, a_n$ with integral coefficients. If, in particular, $a_1, \ldots, a_n$ are integers, then $Q(\xi_1, \ldots, \xi_n)$ is an integer.

We now state

Theorem 15. *Let* $P(\xi) = \xi^n + a_1\xi^{n-1} + \ldots + a_n$ *be an irreducible polynomial with integral coefficients* $a_1, \ldots, a_n$. *If all the zeros of* P *are real, and* Δ *denotes the discriminant of* P, *then we have*

$$\Delta \geq \left(\frac{n^n}{n!} \right)^2 .$$

The proof will be obtained by considering the following linear forms:

$$(11) \qquad y_j = \sum_{k=1}^{n} \xi_j^{n-k} x_k , \quad j = 1, \ldots, n ,$$

where the x_k, $k = 1, \ldots, n$, are arbitrary *integers*, and $\xi_1, \ldots, \xi_n$ are the n distinct (as stated before) zeros of P. The forms y_j, $j = 1, \ldots, n$, are never zero, except in the case $x_1 = x_2 = \ldots = x_n = 0$, for otherwise ξ_j will be a zero of a polynomial, not identically zero, with integral coefficients, and of degree $\leq n - 1$, which contradicts the irreducibility of P.

Therefore the product $y_1 y_2 \ldots y_n$ cannot be zero, unless $x_1 = x_2 = \ldots$ $\ldots = x_n = 0$, and it must be an integer, because it is a symmetric polynomial with integral coefficients in the zeros of P. We know then that

(12) $$|y_1 y_2 \ldots y_n| \geq 1 \, ,$$

provided that the case $x_1 = x_2 = \ldots = x_n = 0$ is excluded.

If we introduce the gauge function

$$f(y) = \frac{1}{n} \sum_{j=1}^{n} |y_j| \, , \quad y = (y_1, \ldots, y_n) \, ,$$

and *define* $\mu = \min f(y)$, where the minimum is taken over all lattice points $y = (y_1, \ldots, y_n) \neq (0, \ldots, 0)$ defined in (11), Theorem 13 states that

(13) $$V \mu^n \leq 2^n D \, ,$$

where D stands for the absolute value of the determinant of (11), and V stands for the volume of the convex body $\{y \,|\, f(y) < 1\}$.

Since V is the volume of the n-dimensional unit octahedron increased in the ratio $n : 1$, we find by the use of (9) that

$$V = \frac{(2n)^n}{n!} \, ,$$

and from (10) we see that $D = \sqrt{\Delta}$. Substituting in (13), we have

(14) $$\sqrt{\Delta} \geq \frac{(\mu n)^n}{n!} \, .$$

Now from the inequality of the arithmetic and geometric means [Lecture II, formula (8)], and (12), it follows that

$$\frac{1}{n} \sum_{j=1}^{n} |y_j| \geq |y_1 y_2 \ldots y_n|^{1/n} \geq 1 \, ,$$

for all admissible $(y_1, \ldots, y_n)$. Hence we have

$$\mu = \min \frac{1}{n} \sum_{j=1}^{n} |y_j| \geq 1 \, ,$$

which, when combined with (14), gives the inequality $\sqrt{\Delta} \geq \frac{n^n}{n!}$, as claimed in the theorem.

As an illustration, take the case $n = 2$, so that Theorem 15 tells us that $\Delta \geq 4$. Let the irreducible polynomial be $P(\xi) = \xi^2 + a\xi + b$. Its discriminant is $\Delta = a^2 - 4b$. Since the zeros of P are real and distinct, $\Delta > 0$. Since a and b are integers, we may have Δ equal to 1, 2, 3, 4 or a larger integer. However if $\Delta = 1$ or 4, the polynomial is reducible. [If $\Delta = 1$, the zeros are $-\frac{a}{2} \pm \frac{1}{2}$, and if $\Delta = 4$, the zeros are $-\frac{a}{2} \pm 1$.] The non-trivial part of our theorem is that Δ cannot be 2 or 3. It is easy to show this directly for $\Delta = a^2 - 4b \equiv a^2$ (mod 4), and any square (of an integer) must be congruent to 0 or 1 (mod 4). Hence $\Delta > 4$, which shows that the bound given by the theorem is not exact. Note that if $a = 1$, $b = -1$, then $\Delta = 5$, and this is the exact lower bound for $n = 2$.

It might be thought that a better bound could be found for Δ, by using a slightly different gauge function. We will show that the bound obtained cannot thus be improved.

It is clear from (13) that the value of the bound will depend on the value of V. Suppose we use the gauge function

$$f_r(y) = \left(\frac{1}{n} \sum_{j=1}^{n} |y_j|^r \right)^{1/r} \quad , \quad r \geq 1 , \quad y = (y_1, \ldots, y_n) .$$

Let $V(r)$ denote the volume of the convex body $\{y | f_r(y) < 1\}$. We shall show that

$$(15) \qquad\qquad V(r) \leq V(1) ,$$

so that the proof of Theorem 15 does not yield a better bound for Δ.
Set

$$f_r(y) = \left(\frac{1}{n} \sum_{j=1}^{n} |y_j|^r \right)^{1/r} \quad , \quad \text{for } r > 0 , \quad y = (y_1, \ldots, y_n) ,$$

and denote by $V(r)$ the volume of the bounded open set $\{y | f_r(y) < 1\}$. Then inequality (15) is a special case of the inequality

$$(15)' \qquad\qquad V(r) \leq V(s) , \quad \text{for } 0 < s < r ,$$

which one can prove by use of Hölder's inequality:

$$(16) \qquad \sum_{j=1}^{n} a_j^p \cdot b_j^{1-p} \leq \left(\sum_{j=1}^{n} a_j \right)^p \cdot \left(\sum_{j=1}^{n} b_j \right)^{1-p} ,$$

for $a_j \geq 0$, $b_j \geq 0$, $j = 1, \ldots, n$, and $0 < p < 1$. [See Lecture II, (13).]
Since $0 < s < r$, we can take

$$p = \frac{s}{r} ; \quad a_j = \frac{1}{n} |y_j|^r , \quad b_j = \frac{1}{n} , \quad j = 1, \ldots, n .$$

Then from (16) we have

(17)
$$\frac{1}{n}\sum_{j=1}^{n}|y_j|^s \leq \left(\frac{1}{n}\sum_{j=1}^{n}|y_j|^r\right)^{s/r},$$

since $\sum_{j=1}^{n} b_j = 1$. Inequality (17) can be written as

$$f_s(y) = \left(\frac{1}{n}\sum_{j=1}^{n}|y_j|^s\right)^{1/s} \leq \left(\frac{1}{n}\sum_{j=1}^{n}|y_j|^r\right)^{1/r} = f_r(y).$$

This shows that if $y \in \mathcal{B}_r = \{y | f_r(y) < 1\}$, then it must also belong to $\mathcal{B}_s$; hence (15)' is proved. Since s can be taken equal to 1, we have finally

$$V(r) \leq V(1), \quad \text{for all } r \geq 1.$$

§3. Successive minima

Let f be an even gauge function on $\mathbb{R}^n$, and let $\mathcal{B}$ denote the convex body $\{x | f(x) < 1\}$. For $\lambda > 0$, let $\lambda\mathcal{B}$ denote the convex body $\{x | f(x) < \lambda\}$. This is obtained by stretching all the coordinates of the points in $\mathcal{B}$ in the ratio $\lambda : 1$. Consider the number of g-points (points all of whose coordinates are integers) inside $\lambda\mathcal{B}$. For $\lambda > 0$, and λ small enough, it is clear that $\lambda\mathcal{B}$ will contain no g-point other than the origin. As λ increases, there will be a value of λ, call it ν_1, such that the origin is the only g-point inside $\nu_1\mathcal{B}$, but such that there exist g-points, different from the origin, on $\partial(\nu_1\mathcal{B})$, the surface of $\nu_1\mathcal{B}$.

Since $\mathcal{B}$ has a centre, there are at least two such points on the surface of $\nu_1\mathcal{B}$, but there may be more. Suppose that *all* g-points on the surface of $\nu_1\mathcal{B}$ can be represented as a linear combination of k_1 linearly independent vectors, $x^{(1)}, x^{(2)}, \ldots, x^{(k_1)}$. [We say that k vectors $x^{(1)}, \ldots, x^{(k)}$ are linearly independent if a linear relation among the vectors, such as $\sum_{j=1}^{k}\lambda_j x^{(j)} = 0$, implies that all the λ_j, $j = 1, \ldots, k$, are zero.]

Now continue expanding $\mathcal{B}$ until we reach a value of λ, ν_2 say, such that on the surface of $\nu_2\mathcal{B}$, but not in its interior, there are g-points (whose vectors are) linearly independent of $x^{(1)}, \ldots, x^{(k_1)}$. Suppose these new g-points can be spanned by k_2 linearly independent vectors. We can continue this process until we have obtained a set of n linearly independent vectors. Then we have $k_1 + k_2 + \ldots + k_j = n$.

Let
$$\mu_1 = \nu_1, \quad \mu_2 = \nu_1, \ldots, \quad \mu_{k_1} = \nu_1,$$
$$\mu_{k_1+1} = \nu_2, \quad \mu_{k_1+2} = \nu_2, \ldots, \quad \mu_{k_1+k_2} = \nu_2,$$
$$\ldots$$

and so on. In this way we have defined uniquely n numbers $\mu_1, \ldots, \mu_n$, and we have also defined n linearly independent vectors $x^{(1)}, \ldots, x^{(n)}$ corresponding to them.

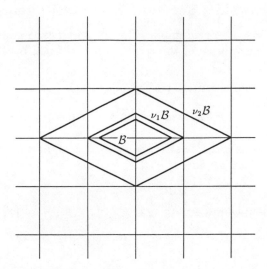

It is clear that the vectors $x^{(1)}, \ldots, x^{(n)}$, are *not* unique, for we can always, for example, replace $x^{(1)}$ by $-x^{(1)}$.

These numbers $\mu_1, \ldots, \mu_n$ are called *successive minima* of the even gauge function f, because they can also be defined as follows:

μ_1 is the minimum value of $f(g)$ over all the g-points $g \neq 0$. [The minimum exists, since for any real number $\alpha > 0$, there are only a finite number of g-points g, such that $f(g) < \alpha$.] If μ_1 is attained for the vector $x^{(1)}$, that is $f(x^{(1)}) = \mu_1$, then define μ_2 as the minimum of $f(g)$ taken over all g-points (whose vectors are) linearly independent of $x^{(1)}$. Let μ_2 be attained for the vector $x^{(2)}$, that is $f(x^{(2)}) = \mu_2$. Then μ_3 is defined as the minimum of $f(g)$ taken over all g-points (whose vectors are) linearly independent of $x^{(1)}$, $x^{(2)}$, and so on.

We will show that this definition of μ_1 agrees with the previous definition of μ_1 as ν_1. Suppose, if possible, that $\nu_1 < \mu_1$. Then there exists a g-point $x^{(1)} \neq 0$ on the surface of $\nu_1 B$, that is $f(x^{(1)}) = \nu_1$. But this contradicts the definition of μ_1 as the minimum of $f(g)$ for *all* g-points $g \neq 0$. If, on the other hand, $\mu_1 < \nu_1$, there exists a g-point $x^{(1)} \neq 0$, such that $f(x^{(1)}) = \mu_1 < \nu_1$, so that $x^{(1)}$ is a g-point in $\nu_1 B$, which contradicts the definition of ν_1 as a value of λ, such that there are no g-points, different from the origin, inside λB. Hence we have $\nu_1 = \mu_1$. The proof is similar for the other μ's, and we see that the two definitions coincide.

§4. Minkowski's Second Theorem (Theorem 16)

Minkowski's First Theorem implies that given an even gauge function f on $\mathbb{R}^n$, if V denotes the volume of the convex body defined by $\{x | f(x) < 1\}$, we have

(18) $$V \mu_1^n \leq 2^n \,,$$

[see Theorem 12], where μ_1 is defined as in § 3. We can generalize it by stating

Theorem 16. *If $\mu_1, \ldots, \mu_n$ denote the successive minima of an even gauge function f on $\mathbb{R}^n$, then we have*

(19) $$V \mu_1 \mu_2 \ldots \mu_n \leq 2^n \,,$$

where V denotes the volume of the convex body defined by $\{x | f(x) < 1\}$.

Note that since $\mu_1 \leq \mu_2 \leq \ldots \leq \mu_n$, inequality (19) is stronger than inequality (18).

We can illustrate the content of Theorem 16 by considering the case where the coordinate vectors are the *minimizing vectors* $x^{(1)}, \ldots, x^{(n)}$, that is vectors (corresponding to points) at which f attains the successive minima. [See § 3, $f(x^{(1)}) = \mu_1$, $f(x^{(2)}) = \mu_2, \ldots$.] If we multiply all the coordinates by μ_1, the convex body $\{x | f(x) < 1\}$ goes over into the convex body $\{x | f(x) < \mu_1\}$, which has the volume $V \mu_1^n$, and which has no g-point inside it different from the origin. Minkowski's First Theorem tells us that

$$V \mu_1^n \leq 2^n \,.$$

Suppose now we multiply the first coordinate by μ_1, the second by μ_2, and so on. We get again a convex body $\mathcal{C}$ with volume $V \mu_1 \mu_2 \ldots \mu_n$. If this convex body contains no g-point different from the origin, then $V \mu_1 \mu_2 \ldots \mu_n \leq 2^n$. In short, instead of expanding each coordinate by the same amount, we expand the kth coordinate in the ratio $\mu_k : 1$, or until it reaches its first g-point, and then apply Minkowki's First Theorem. [But the trouble is that the transformed convex body may indeed contain a g-point.]

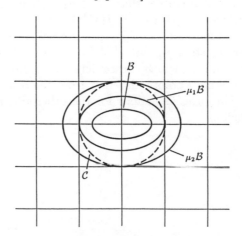

Lecture IV

§1. A possible method of proof

Let f be an even gauge function on $\mathbb{R}^n$, and $\mathcal{B}$ the convex body defined by $\{x|f(x) < 1\}$. Let $\mu_1, \ldots, \mu_n$ be the *successive minima* of f, and let the minima be attained at the vectors $x^{(1)}, \ldots, x^{(n)}$, called the *minimizing vectors*. Another way of looking at this fact is as follows:

If x is any *integral vector* [also called sometimes a *g-point*, or *g-vector*] which is linearly independent of the first $k - 1$ minimizing vectors

$$(1) \qquad\qquad x^{(1)}, \ldots, x^{(k-1)} \, ,$$

then

$$(2) \qquad\qquad f(x) \geq f(x^{(k)}) = \mu_k \, .$$

This statement holds for $k = 1, \ldots, n$. For $k = 1$ it means that if x is not the zero vector, then

$$f(x) \geq \mu_1 \, .$$

Since the n minimizing vectors are linearly independent, *any vector x in $\mathbb{R}^n$* can be written as

$$(3) \qquad\qquad x = \lambda_1 x^{(1)} + \ldots + \lambda_n x^{(n)} \, ,$$

where the λ_j, for $j = 1, \ldots, n$, are real numbers. Now we perform the following transformation:

Multiply the first minimizing vector $x^{(1)}$ by μ_1; the second, $x^{(2)}$, by μ_2; and so on. Then to every vector x in $\mathcal{B}$ will correspond a vector x', such that

$$(4) \qquad\qquad x' = \lambda_1 \mu_1 x^{(1)} + \ldots + \lambda_n \mu_n x^{(n)} \, .$$

Graphically, what we have done is to take the parallelepiped formed by the minimizing vectors, and expand the kth vector in it in the ratio $\mu_k : 1$. It is clear that under this transformation, straight lines go over into straight lines, so that the convex body $\mathcal{B}$ goes over into a convex body $\mathcal{B}'$.

If V is the volume of $\mathcal{B}$, then $V \mu_1 \ldots \mu_n$ will be the volume of $\mathcal{B}'$. *Suppose the volume of $\mathcal{B}'$ is greater than 2^n.* By Minkowski's First Theorem, there exists in $\mathcal{B}'$ a *g*-point, say g', different from the origin. We can write

$$(5) \qquad\qquad g' = \lambda_1' x^{(1)} + \ldots + \lambda_n' x^{(n)} \, .$$

Since the points of $\mathcal{B}'$ were obtained from the points of $\mathcal{B}$ by the transformation defined by (4), we find that the point $g' \in \mathcal{B}'$ corresponds to some point $x \in \mathcal{B}$ [which, in general, is *not* a g-point], given by

$$(6) \qquad\qquad x = \lambda_1 x^{(1)} + \ldots + \lambda_n x^{(n)} ,$$

where

$$\lambda_1' = \mu_1 \lambda_1, \ldots, \lambda_n' = \mu_n \lambda_n .$$

Let $\lambda_k \neq 0$, and $\lambda_{k+1} = \ldots = \lambda_n = 0$, so that g' is linearly dependent on $x^{(1)}, \ldots, x^{(k)}$, where k may be *any* integer from 1 to n. We want to evaluate $f(g')$.

Note that g' can be written as

$$
(7) \quad
\begin{aligned}
g' = \mu_1 \left(\lambda_1 x^{(1)} + \ldots + \lambda_k x^{(k)} \right) + (\mu_2 - \mu_1) \left(\lambda_2 x^{(2)} + \ldots + \lambda_k x^{(k)} \right) \\
+ \ldots + (\mu_k - \mu_{k-1}) \lambda_k x^{(k)} .
\end{aligned}
$$

On using the convexity property of the gauge function f [Lecture I, Theorem 6], we have

$$
(8) \quad
\begin{aligned}
f(g') \leq \mu_1 f\left(\lambda_1 x^{(1)} + \ldots + \lambda_k x^{(k)} \right) + (\mu_2 - \mu_1) f\left(\lambda_2 x^{(2)} + \ldots + \lambda_k x^{(k)} \right) \\
+ \ldots + (\mu_k - \mu_{k-1}) f\left(\lambda_k x^{(k)} \right) .
\end{aligned}
$$

We know that the point $x = \lambda_1 x^{(1)} + \ldots + \lambda_k x^{(k)}$ belongs to $\mathcal{B}$, so that $f(\lambda_1 x^{(1)} + \ldots + \lambda_k x^{(k)}) < 1$. If we could be sure that the other vectors $\lambda_j x^{(j)} + \ldots + \lambda_k x^{(k)}$, $2 \leq j \leq k$, are also in $\mathcal{B}$, we would have from (8)

$$f(g') < \mu_1 + (\mu_2 - \mu_1) + \ldots + (\mu_k - \mu_{k-1}) = \mu_k .$$

This contradicts, however, the minimizing property of the vectors $x^{(1)}, \ldots, x^{(k)}$, as can be seen from (1) and (2), if we observe that g' is linearly independent of $x^{(1)}, \ldots, x^{(k-1)}$, since $\lambda_k \neq 0$. Therefore our assumption that the volume of $\mathcal{B}'$ is greater than 2^n must be wrong, and Minkowski's Second Theorem would be proved.

There still remains the question, whether the vectors

$$\lambda_j x^{(j)} + \ldots + \lambda_k x^{(k)} , \quad 2 \leq j \leq k ,$$

represent points in $\mathcal{B}$, if $\lambda_1 x^{(1)} + \ldots + \lambda_k x^{(k)}$ represents a point in $\mathcal{B}$.

The diagram shows that even in the simple case $n = 2$, and $f(x) = f(x_1, x_2) = (2x_1^2 - x_1 x_2 + 3x_2^2)^{1/2}$, the answer is in the negative. Here the successive minima are: $\mu_1 = \sqrt{2}$, $\mu_2 = \sqrt{3}$, which correspond to the minimizing vectors $(1,0)$ and $(0,1)$. Let $(\tau_1, 0)$ and $(0, \tau_2)$ be the points of intersection of the ellipse $\{x | f(x) = 1\}$ with the positive x_1-axis, and x_2-axis, respectively. Then there exist points (x_1', x_2'), such that $x_1' > \tau_1$, or $x_2' > \tau_2$, for which $x_1'(1,0) + x_2'(0,1) \in \mathcal{B}$, but the points $(x_1', 0)$ or $(0, x_2')$ do not belong to $\mathcal{B}$.

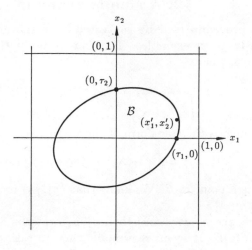

This fact causes our method to fail, but the proof can still be saved if we use a more complicated transformation of $\mathcal{B}$. It is clear that we must change (7) so that the vector g' will be expressed as a linear combination of vectors in $\mathcal{B}$.

Let $x \in \mathcal{B}$, with $x = \lambda_1 x^{(1)} + \ldots + \lambda_n x^{(n)}$, as in (3). Let x' be the image of x under the transformation which we are seeking to construct. Let us write

$$x' = \mu_1 y^{(1)} + (\mu_2 - \mu_1) y^{(2)} + \ldots + (\mu_n - \mu_{n-1}) y^{(n)} \, ,$$

where $y^{(1)}, \ldots, y^{(n)}$ are vectors in $\mathcal{B}$. We can then express $y^{(1)}, \ldots, y^{(n)}$ as linear combinations of the minimizing vectors $x^{(1)}, \ldots, x^{(n)}$, with coefficients depending on $\lambda_1, \ldots, \lambda_n$. That is to say,

$$y^{(k)} = \sum_{j=1}^{n} \alpha_{kj}(\lambda_1, \ldots, \lambda_n) x^{(j)} \, , \quad \text{for } k = 1, \ldots, n \ .$$

Since the volume of the image of $\mathcal{B}$ must be $V\mu_1 \ldots \mu_n$, where V is the volume of $\mathcal{B}$, we want the coefficient functions of $y^{(k)}$ to be independent of $\lambda_1, \ldots, \lambda_{k-1}$, that is

$$\alpha_{kj}(\lambda_1, \ldots, \lambda_n) \equiv \alpha_{kj}(\lambda_k, \ldots \lambda_n) \, , \quad \text{for } j, k = 1, \ldots, n \ .$$

Secondly, in order to make use of property (2) of the minimizing vectors, we want that

$$\alpha_{kk}(\lambda_k, \lambda_{k+1}, \ldots, \lambda_n) \neq \alpha_{kk}(\widetilde{\lambda}_k, \lambda_{k+1}, \ldots, \lambda_n) \, ,$$

for $\lambda_k \neq \widetilde{\lambda}_k$, $k = 1, \ldots, n$. Our object will be to construct a transformation which secures these conditions. Note how these are met in (7). The jth vector there was $\lambda_j x^{(j)} + \ldots + \lambda_k x^{(k)}$, but, as we saw, this may not correspond to a point in $\mathcal{B}$.

§ 2. A simple example

A possible procedure will be suggested by a more careful consideration of the previously given example. The convex body $\mathcal{B}$ in $\mathbb{R}^2$ is defined by $\mathcal{B} = \{x \mid f(x) < 1\}$, where

$$f(x) = f(x_1, x_2) = (2x_1^2 - x_1 x_2 + 3x_2^2)^{1/2} \ .$$

We want to transform $\mathcal{B}$ into a bounded open set $\mathcal{C}$, such that for each point (λ_1', λ_2') in $\mathcal{C}$, which is the transform of the point (λ_1, λ_2) in $\mathcal{B}$, we can write

(9) $$(\lambda_1', \lambda_2') = \mu_1(\lambda_1, \lambda_2) + (\mu_2 - \mu_1)(p_1, p_2) \ ,$$

where (p_1, p_2) is a point in $\mathcal{B}$. We know that (p_1, p_2) must depend on λ_2, and cannot depend on λ_1.

Suppose we take λ_2 constant; then p_1, p_2 are constant, and yet the point (p_1, p_2) belongs to $\mathcal{B}$. It seems reasonable then to take (p_1, p_2) as some point on the intersection $\mathcal{E}$ of $\mathcal{B}$ with the straight line $x_2 = \lambda_2$. We set

(10) $$p_1 = c_1(\lambda_2) \ , \quad p_2 = c_2(\lambda_2) \ ,$$

where $c_1(\lambda_2)$ and $c_2(\lambda_2)$ are the coordinates of the centre of gravity of the intersection $\mathcal{E}$. [The only properties of the centre of gravity we shall use are that its position varies continuously with λ_2, and that the centre of gravity of a convex body lies in that convex body.]

We can now define the transformation of $\mathcal{B}$. Because of (9) and (10) we have to set

(11) $$\begin{aligned} \lambda_1' &= \mu_1 \lambda_1 + (\mu_2 - \mu_1) c_1(\lambda_2) \ , \\ \lambda_2' &= \mu_1 \lambda_2 + (\mu_2 - \mu_1) c_2(\lambda_2) \ , \end{aligned}$$

and note that since $c_2(\lambda_2) = \lambda_2$, we really have $\lambda_2' = \mu_2 \lambda_2$. This transformation may be considered as a combination of a dilatation and a shear transformation. It can be seen that, under this transformation, $\mathcal{B}$ goes over into an elliptic disc $\mathcal{C}$ with volume $V\mu_1\mu_2$, where V is the volume of $\mathcal{B}$ $\left[\mu_1 = \sqrt{2} \, , \mu_2 = \sqrt{3} \right.$ (see § 1), $V = \pi\sqrt{\frac{4}{23}} \, , \mathcal{C} = \left\{ (x_1, x_2) \mid (x_1^2 - \frac{1}{2}x_1 x_2 + \frac{49}{48}x_2^2)^{1/2} < 1 \right\} \Big]$.

Now the proof proceeds as before. If the volume of $\mathcal{C}$ is greater than 4, there exists a g-point $g \neq 0$ in $\mathcal{C}$, and we get a contradiction. This shows that $V\mu_1\mu_2 \leq 4$.

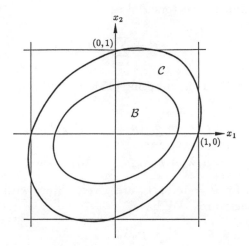

§3. A complicated transformation

In the general case, we proceed similarly. First we set up the transformation of $\mathcal{B}$ into $\mathbb{R}^n$.

Let $x \in \mathcal{B}$, and suppose that

$$(12) \qquad x = \sum_{j=1}^{n} \lambda_j x^{(j)} \, ,$$

$x^{(j)}$, $j = 1, \ldots, n$, being the minimizing vectors. Consider the set of all points in $\mathbb{R}^n$ having the first k coordinates, with respect to the basis $x^{(1)}, \ldots, x^{(n)}$, as variables, and the last $n-k$ fixed and equal to $\lambda_{k+1}, \ldots, \lambda_n$. These points determine a k-dimensional linear manifold $\mathcal{L} = \mathcal{L}(\lambda_{k+1}, \ldots, \lambda_n)$ of $\mathbb{R}^n$. Since $x \in \mathcal{B}$, the intersection with $\mathcal{B}$ of this linear manifold is not empty. Let $b(\lambda_{k+1}, \ldots, \lambda_n)$ denote the centre of gravity of this intersection. Since $\mathcal{B}$ and $\mathcal{L}$ are convex, $b(\lambda_{k+1}, \ldots, \lambda_n) \in \mathcal{B} \cap \mathcal{L}$. Since $b(\lambda_{k+1}, \ldots, \lambda_n) \in \mathcal{L}$, its last $n - k$ coordinates are constant, and we can write

$$(13) \qquad b(\lambda_{k+1}, \ldots, \lambda_n) = \sum_{j=1}^{k} c_j(\lambda_{k+1}, \ldots, \lambda_n) x^{(j)} + \sum_{j=k+1}^{n} \lambda_j x^{(j)} \, .$$

In particular, for $k = 0$ we have just $b(\lambda_1, \ldots, \lambda_n) = x$.

Consider the transformation $x \mapsto x'$ defined by

$$(14) \quad x' = \mu_1 b(\lambda_1, \ldots, \lambda_n) + (\mu_2 - \mu_1) b(\lambda_2, \ldots, \lambda_n) + \ldots + (\mu_n - \mu_{n-1}) b(\lambda_n) \, ,$$

where x is given by (12). This is a transformation of $\mathcal{B}$ into $\mathbb{R}^n$. Note that x' is uniquely determined by x. We show that, conversely, given any x' in the range of the transformation, it will determine x uniquely. First let x' be expressed in terms of the minimizing vectors $x^{(k)}$ as

$$(15) \qquad x' = \sum_{k=1}^{n} \lambda'_k x^{(k)} \, .$$

Comparing (15) with (14), and using (13), we find that

$$(16) \qquad \begin{aligned} \lambda'_k &= \mu_k \lambda_k + \sum_{j=k}^{n-1} (\mu_{j+1} - \mu_j) c_k(\lambda_{j+1}, \ldots, \lambda_n) \\ &= \mu_k \lambda_k + h(\lambda_{k+1}, \ldots, \lambda_n) \, , \quad \text{say} \, . \end{aligned}$$

Suppose the λ'_k, for $k = 1, \ldots, n$, are given. Start with the last equation in (16) and work backwards. We have

$$\lambda'_n = \mu_n \lambda_n \, ,$$
$$\lambda'_{n-1} = \mu_{n-1} \lambda_{n-1} + h(\lambda_n) \, ,$$

and so on. We find λ_n from the first equation, then λ_{n-1} from the next, and so on. This shows that the λ_j are uniquely determined by the λ'_k, where $j, k = 1, \ldots, n$. It follows that the transformation in (14) is one-to-one.

§4. Volume of the transformed body

[A bounded open set in $\mathbb{R}^n$ is called a *body*.] If we apply transformation (14) to the convex body $\mathcal{B}$ we started with, we get a bounded open set $\mathcal{C}$. In general, $\mathcal{C}$ will not be convex. We wish to show that the volume of $\mathcal{C}$ is $V\mu_1\mu_2 \ldots \mu_n$, where V denotes the volume of $\mathcal{B}$.

If the functions $h : (\lambda_{k+1}, \ldots, \lambda_n) \mapsto h(\lambda_{k+1}, \ldots, \lambda_n)$ are zero for all $k = 1, \ldots, n$, then the transformation in (16) is just a dilatation of the kth minimizing vector by μ_k, and the volume of $\mathcal{C}$ will be $V\mu_1 \ldots \mu_n$.

Suppose the functions h are not zero. Then

$$(17) \qquad \text{volume of } \mathcal{C} = \Theta \int \ldots \int d\lambda'_1 \ldots d\lambda'_n \, ,$$

where the integration is over all values of $\lambda'_1, \ldots, \lambda'_n$, such that the point x' given by (15) is a point in $\mathcal{C}$. Here Θ denotes the absolute value of the determinant of the linear transformation from the original coordinate vectors to the minimizing vectors.

Consider first the case where (16) reduces to

$$(18) \qquad \begin{aligned} \lambda'_1 &= \mu_1 \lambda_1 + h(\lambda_2, \ldots, \lambda_n) \, , \\ \lambda'_2 &= \mu_2 \lambda_2 \, , \\ &\;\;\vdots \\ \lambda'_n &= \mu_n \lambda_n \, . \end{aligned}$$

Suppose the interval of integration for λ_1' in (17) is from $f_1(\lambda_2,\ldots,\lambda_n)$ to $f_2(\lambda_2,\ldots,\lambda_n)$. [It is an interval since it is the image of an interval under a continuous function.] By (17) and (18) we then have

$$\text{volume of } C = \Theta \int \cdots \int \left[\int_{(f_1-h)/\mu_1}^{(f_2-h)/\mu_1} \mu_1 \, d\lambda_1 \right] d\lambda_2' \ldots d\lambda_n' \; ,$$

where $h \equiv h(\lambda_2,\ldots,\lambda_n)$ is only an abbreviation in writing. Carrying out the integration relative to $d\lambda_1$, we get

$$\text{volume of } C = \Theta \int \cdots \int (f_2 - f_1)(\lambda_2,\ldots,\lambda_n) \, d\lambda_2' \ldots d\lambda_n'$$

$$= \Theta \int \cdots \int d\lambda_1' \ldots d\lambda_n' \; ,$$

and we see that this is the same *as if* $h(\lambda_2,\ldots,\lambda_n)$ had been zero.

Geometrically, what we have shown is that the volume of C is unchanged if we translate it parallel to some of the coordinate planes with respect to the basis of minimizing vectors. The same thing can be done step-wise for the more complicated transformation in (16). We first show that the volume of C is independent of $h(\lambda_2,\ldots,\lambda_n)$, and then that it is independent of $h(\lambda_3,\ldots,\lambda_n)$, and so on. This proves that the volume of C is the same as the volume obtained under a simple dilatation: $\lambda_k' = \mu_k \lambda_k$, $k = 1,\ldots,n$, namely $V\mu_1 \ldots \mu_n$.

§5. Proof of Theorem 16 (Minkowski's Second Theorem)

Assume that $V\mu_1 \ldots \mu_n > 2^n$. We shall show that this leads to a contradiction. [Recall that V is the volume of the convex body $B = \{x | f(x) < 1\}$, where f is an even gauge function given on $\mathbb{R}^n$, and C is the bounded open set in $\mathbb{R}^n$ obtained from B by the transformation given in (14), the volume of C being $V\mu_1 \ldots \mu_n$.] Consider the bounded open set $\frac{1}{2}C = \{x | 2x \in C\}$. It has a volume greater than 1, by assumption. By the lemma used in the proof of Minkowski's First Theorem, there exist two *distinct* points $\frac{1}{2}y, \frac{1}{2}\tilde{y}$ in $\frac{1}{2}C$, such that the difference $\frac{1}{2}\{y - \tilde{y}\}$ is a g-point. The points $y, \tilde{y}$ lie in C. Let $x, \tilde{x}$ be the points in B whose images are $y, \tilde{y}$ in C. Suppose that

$$x = \sum_{j=1}^{n} \lambda_j x^{(j)} \; , \qquad \tilde{x} = \sum_{j=1}^{n} \tilde{\lambda}_j x^{(j)} \; .$$

Since the transformation from B to C is one-to-one, it follows that $x \neq \tilde{x}$, so that $\lambda_k \neq \tilde{\lambda}_k$, for some integer k, $1 \leq k \leq n$, but $\lambda_{k+1} = \tilde{\lambda}_{k+1},\ldots,\lambda_n = \tilde{\lambda}_n$. Then we shall show that $f(\frac{1}{2}\{y - \tilde{y}\}) < \mu_k$. This contradicts our definition of the minimizing vectors, because of (2), since $\frac{1}{2}\{y - \tilde{y}\}$ is an integral vector linearly independent of $x^{(1)},\ldots,x^{(k-1)}$.

From (14) we have

$$\frac{1}{2}\{y - \widetilde{y}\} = \mu_1 \cdot \frac{1}{2}\{b(\lambda_1, \ldots, \lambda_n) - b(\widetilde{\lambda}_1, \ldots, \widetilde{\lambda}_n)\}$$

(19)
$$+ (\mu_2 - \mu_1) \cdot \frac{1}{2}\{b(\lambda_2, \ldots, \lambda_n) - b(\widetilde{\lambda}_2, \ldots, \widetilde{\lambda}_n)\}$$

$$+ \ldots + (\mu_k - \mu_{k-1}) \cdot \frac{1}{2}\{b(\lambda_k, \ldots, \lambda_n) - b(\widetilde{\lambda}_k, \ldots, \widetilde{\lambda}_n)\} \; .$$

We can stop at $\mu_k - \mu_{k-1}$, since all the coordinates after the kth are the same.

All the points in (19) belong to $\mathcal{B}$, namely

$$\frac{1}{2}\{b(\lambda_j, \ldots, \lambda_n) - b(\widetilde{\lambda}_j, \ldots, \widetilde{\lambda}_n)\} \; , \quad j = 1, \ldots, k \; ,$$

since $\mathcal{B}$ is a convex body with a centre. Because of the convexity property of the gauge function [Lecture I, Theorem 6], we have

$$f\left(\frac{1}{2}\{y - \widetilde{y}\}\right) \le \mu_1 f\left(\frac{1}{2}\{b(\lambda_1, \ldots, \lambda_n) - b(\widetilde{\lambda}_1, \ldots, \widetilde{\lambda}_n)\}\right)$$

$$+ (\mu_2 - \mu_1)f\left(\frac{1}{2}\{b(\lambda_2, \ldots, \lambda_n) - b(\widetilde{\lambda}_2, \ldots, \widetilde{\lambda}_n)\}\right)$$

$$+ \ldots + (\mu_k - \mu_{k-1})f\left(\frac{1}{2}\{b(\lambda_k, \ldots, \lambda_n) - b(\widetilde{\lambda}_k, \ldots, \widetilde{\lambda}_n)\}\right) \; .$$

Since $f(z) < 1$, for $z \in \mathcal{B}$, we have

$$f\left(\frac{1}{2}\{y - \widetilde{y}\}\right) < \mu_1 + (\mu_2 - \mu_1) + \ldots + (\mu_k - \mu_{k-1}) = \mu_k \; ,$$

the desired contradiction.

This shows that we must have: $V\mu_1 \ldots \mu_n \le 2^n$.

We now prove an inequality in the opposite direction:

$$2^n \frac{\Theta}{n!} \le V\mu_1 \ldots \mu_n \; ,$$

where Θ is defined as before.

The proof is simple. We know that $x^{(k)}$ is a point on the surface of $\mu_k\mathcal{B}$, so that $\frac{1}{\mu_k}x^{(k)}$ is a point on the surface of $\mathcal{B}$. In this way, the n minimizing vectors determine n points on the surface of $\mathcal{B}$, but since $\mathcal{B}$ has a centre, the points $-\frac{1}{\mu_k}x^{(k)}$ must also lie on the surface of $\mathcal{B}$. These $2n$ points are the vertices of an open n-dimensional octahedron. If we refer to Lecture III, we find that its volume is

$$\frac{2^n \Theta}{n!\mu_1 \ldots \mu_n} \; .$$

Since $\mathcal{B}$ is convex, it must contain this octahedron, and therefore we have

$$\frac{2^n \Theta}{n!\mu_1 \ldots \mu_n} \le V \; .$$

This completes the consideration of Minkowski's Second Theorem.

Chapter II

Linear Inequalities

Lectures V to IX

Lecture V

§1. Vector groups

The main aim of this Chapter will be the question of solving linear equations approximately by means of integers. The ideas developed will then be used in the study of the periods of real functions and of analytic functions.

The discussion can be very much simplified by the use of the concept of *vector groups* or *modules*. A subset G of vectors in n-dimensional real Euclidean space is called a *vector group* or *module*, if it contains at least one element, and if whenever x and y belong to G, then $x - y$ also belongs to G.

If x belongs to G, then from the definition $x - x = 0$ belongs to G, so that every vector group contains the zero vector 0. Similarly if x belongs to G, then $0 - x = -x$ belongs to G, and so if x and y belong to G, then $x - (-y) = x + y$ belongs to G. From this it follows that $x + x = 2x$ belongs to G, and also that μx belongs to G, where μ is an integer.

Generally, if $x^{(1)}, \ldots, x^{(k)}$ are vectors belonging to a vector group G, then all vectors of the form

$$(1) \qquad\qquad g_1 x^{(1)} + \ldots + g_k x^{(k)} \ ,$$

where $g_1, \ldots, g_k$ are integers, belong to G. If G contains no other vectors, G is said to be *generated* by $x^{(1)}, \ldots, x^{(k)}$.

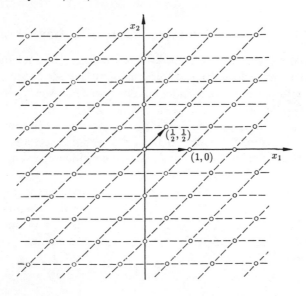

As an illustration consider two-dimensional Euclidean space, and let the vector group G be generated by $x^{(1)} = (1,0)$, $x^{(2)} = (0,1)$, $x^{(3)} = (\frac{1}{2}, \frac{1}{2})$. Then G consists of all integral vectors and all vectors both of whose coordinates are halves of odd integers. Note that in this case G can also be considered as a lattice formed from the vectors $(1,0)$ and $(\frac{1}{2}, \frac{1}{2})$. Such a representation of G as a lattice may not be possible in all cases; for example, G might be generated by the vectors $(1,0)$ and $(\pi,0)$. [In the next lecture we show that this group is dense on the x_1-axis.] However we shall prove

Theorem 17. *If the vector group G does not contain vectors of arbitrarily small positive length, then there exists a finite number of vectors in G,*

$$x^{(1)}, x^{(2)}, \ldots, x^{(r)} \ , \quad say,$$

such that every vector x in G can be written, in exactly one way, as

$$(2) \qquad\qquad x = g_1 x^{(1)} + g_2 x^{(2)} + \ldots + g_r x^{(r)} \ ,$$

where $g_1, \ldots, g_r$ are integers.

In this case G is called a *discrete vector group*. It follows that a discrete vector group is a *lattice* [in a linear manifold $\mathbb{R}^r$ through the origin of $\mathbb{R}^n$ (cf. Lecture II, § 5)], and the vectors $x^{(1)}, \ldots, x^{(r)}$ are called a *basis*, sometimes an *integral basis*, for the lattice G.

We may state the condition that G does not contain vectors of arbitrarily small positive length in the following way: there exists a real number $\epsilon > 0$, such that if the length of $x \in G$ is less than ϵ, then x must be the zero vector.

It now follows that a *discrete vector group* G is a closed subset of $\mathbb{R}^n$. For, suppose there exists a sequence of vectors $y^{(1)}, y^{(2)}, y^{(3)}, \ldots$ in G, such that $y^{(k)} \to y$, as $k \to \infty$. Then for any $\epsilon > 0$, there exists a number $k = k(\epsilon)$, such that $|y^{(j)} - y^{(k)}| < \epsilon$, for $j > k$. By our condition on G, this inequality implies that $y^{(j)} = y^{(k)}$ for all sufficiently large j and k.

§2. Construction of a basis

In any vector group there are at most n linearly independent vectors, because the vectors belong to an n-dimensional Euclidean space. We call r, the maximum number of linearly independent (over $\mathbb{R}$) vectors belonging to a vector group G, the *rank* of G. It is clear that $0 \le r \le n$.

Note that r linearly independent vectors in a vector group G of rank r may be neither a basis for G (in case G is a lattice) nor a set of elements of G which can generate G. For example, in the lattice considered in § 1, the vectors $(\frac{5}{2}, \frac{5}{2})$ and $(2,0)$ are linearly independent, but the vector $(1,0)$ cannot be represented as a linear combination, with integral coefficients, of $(\frac{5}{2}, \frac{5}{2})$ and $(2,0)$. If, on the other hand, $G = \{\lambda(1,0) + \mu(\sqrt{2},0) | \lambda, \mu \text{ integers}\}$, then G is a vector group generated by $(1,0)$ and $(\sqrt{2},0)$. The rank of G is 1, but since G is, of course, not discrete, G cannot be generated by a single element.

Nevertheless we prove

Theorem 18. *Given any r linearly independent vectors*

$$y^{(1)}, \ldots, y^{(r)}$$

belonging to a discrete vector group (or lattice) G of rank r, there exists a basis $x^{(1)}, \ldots, x^{(r)}$ for G defined as follows:

$$x^{(1)} = c_1 y^{(1)} ,$$

$$x^{(2)} = c_{21} y^{(1)} + c_2 y^{(2)} ,$$

$$\vdots$$

$$x^{(r)} = c_{r1} y^{(1)} + \ldots + c_{r,r-1} y^{(r-1)} + c_r y^{(r)} ,$$

where $c_k > 0$, $1 \le k \le r$. [It will be seen later in § 5, at the beginning of the proof of Theorem 20, that in fact $\frac{1}{c_k} =$ a positive integer for $1 \le k \le r$, and $c_{kj} \in \mathbb{Q}$, the field of rational numbers, for $1 \le j < k \le r$.]

Theorems 17 and 18 will be proved together. The idea of the proof is illustrated in the following example: let the lattice G be generated by $(1,0),(0,1)$, and $(\frac{1}{2},\frac{1}{2})$. Let $y^{(1)} = (\frac{5}{2},\frac{5}{2})$, $y^{(2)} = (-3,9)$. We want to find a basis for G which is of the form stated in Theorem 18.

We first consider all positive values of λ_1, for which $\lambda_1 y^{(1)}$ belongs to G. The smallest such value is $\frac{1}{5}$, and all the other such values are positive integral multiples of it. Now consider the set of all pairs of real numbers (λ_2, λ_3), with $\lambda_2 \ge 0$, $\lambda_3 > 0$, and with the property that $\lambda_2 y^{(1)} + \lambda_3 y^{(2)}$ belongs to G. It is clear that this set contains a pair (λ_2, λ_3), where λ_3 is smallest, namely $(\lambda_2, \lambda_3) = (\frac{1}{10}, \frac{1}{12})$, and that $\lambda_2 y^{(1)} + \lambda_3 y^{(2)} = (0,1)$. The vectors $x^{(1)} = (\frac{1}{2},\frac{1}{2})$, and $x^{(2)} = (0,1)$ obviously form a basis for G, which is of the desired form:

$$x^{(1)} = \frac{1}{5} y^{(1)} , \quad x^{(2)} = \frac{1}{10} y^{(1)} + \frac{1}{12} y^{(2)} .$$

The method of proof in the general case is the same.

Proof of Theorem 18. Let k be an integer, $1 \le k \le r$, and let $\lambda_1, \ldots, \lambda_k$ be non-negative real numbers. Consider vectors of the form

(3) $$z = \lambda_1 y^{(1)} + \ldots + \lambda_k y^{(k)} .$$

What are the values of $\lambda_1, \ldots, \lambda_k$ for which z belongs to G? It is clear that if the λ's are all integers, z belongs to G, but it is possible that the λ's may attain other values.

We will show that the positive values of λ_k have an infimum (greatest lower bound) that is actually attained for some vector z in G. To begin with, let the choice of the λ's be restricted by the following inequalities:

(4)
$$0 \le \lambda_j < 1 , \quad \text{for } j = 1, 2, \ldots, k-1 ,$$
$$0 < \lambda_k .$$

There will be elements z of this type in G; for example, take $\lambda_1 = \lambda_2 = \ldots$
$\ldots = \lambda_{k-1} = 0$, $\lambda_k = $ a positive integer. In the set of all z's belonging to
G which fulfil (4), the values of λ_k will have an infimum. If this infimum is
not attained, there would exist an infinite number of distinct vectors z in G,
such that the value of λ_k for them tends to that infimum. Since all the other
coordinates are bounded, we can pick a subsequence for which all the coordinates converge to a limit, and therefore the corresponding vectors converge to
a limiting vector. Such a fact contradicts the discreteness of the vector group
G, and therefore there must exist a vector $x^{(k)}$ in G, such that

$$(5) \qquad x^{(k)} = c_{k1}y^{(1)} + \ldots + c_{k,k-1}y^{(k-1)} + c_k y^{(k)} \, ,$$

where c_k is the minimum value of λ_k. Note that since $\lambda_k > 0$ we have $c_k > 0$.

This construction may be performed for every value of k, from 1 to r both
inclusive. We obtain r vectors $x^{(1)}, \ldots, x^{(r)}$, each of the form (5). We now
prove that these vectors form a basis for G.

First, by construction, $x^{(1)}, \ldots, x^{(r)}$ belong to G. Secondly they are linearly
independent, because if we had a linear relation among the $x^{(j)}$, $j = 1, \ldots, r$,
for example $\sum_{j=1}^s \lambda_j x^{(j)} = 0$, $1 \le s \le r$, then rewriting it in terms of the $y^{(j)}$,
$j = 1, \ldots, r$, we get

$$\sum_{j=1}^{s-1} \mu_j y^{(j)} + \lambda_s c_s y^{(s)} = 0 \, ,$$

where the μ_j, $j = 1, \ldots, s-1$, are linear combinations of $\lambda_1, \ldots, \lambda_{s-1}$. Since
$y^{(s)}$ is linearly independent of $y^{(1)}, \ldots, y^{(s-1)}$, we must have $\lambda_s = 0$. Similarly
it follows that $\lambda_{s-1} = \lambda_{s-2} = \ldots = \lambda_1 = 0$, which shows that the $x^{(j)}$,
$j = 1, \ldots, r$, are linearly independent.

The only remaining point to be proved is that the $x^{(j)}$, $j = 1, \ldots, r$, form
a basis for G. Since they are linearly independent, we know that any vector x
in G can be written as

$$(6) \qquad x = \sum_{j=1}^r \nu_j x^{(j)} \, ,$$

where the ν's are some real numbers. We have to prove that they are integers.
Suppose that for a certain x in G, *not all* ν's are integers. There would then
exist a first ν_k, counting backwards, which is not an integer, that is

$$x = \sum_{j=1}^k \nu_j x^{(j)} + \nu_{k+1} x^{(k+1)} + \ldots + \nu_r x^{(r)} \, ,$$

where $\nu_{k+1}, \nu_{k+2}, \ldots, \nu_r$ are all integers, and ν_k is *not* an integer.

Since the vectors $x^{(k+1)}, \ldots, x^{(r)}$ belong to G, we have $\sum_{j=k+1}^r \nu_j x^{(j)}$ also
in G. It follows that the vector

$$(7) \qquad x - \sum_{j=k+1}^r \nu_j x^{(j)} = \sum_{j=1}^k \nu_j x^{(j)}$$

belongs to G. Let

$$\nu_k = h_k + r_k \, ,$$

where h_k is an integer, and $0 < r_k < 1$. Subtracting the vector $h_k x^{(k)}$, which belongs to G, from the one defined in (7), we get a vector of the form

$$\sum_{j=1}^{k-1} \nu_j x^{(j)} + \nu_k^* x^{(k)} \, , \qquad \text{where } 0 < \nu_k^* < 1 \, ,$$

which is in G. Substituting for the $x^{(j)}$ in terms of $y^{(j)}$, as given by (5), we get a vector in G of the form

$$(8) \qquad \sum_{j=1}^{k-1} \nu_j' y^{(j)} + \nu_k^* \cdot c_k y^{(k)} \, , \qquad \nu_1', \ldots, \nu_{k-1}' \text{ real.}$$

For $j = 1, \ldots, k-1$, let $\nu_j' = h_j + r_j$, where h_j is an integer, and $0 \le r_j < 1$. Just as before, $\sum_{j=1}^{k-1} h_j y^{(j)}$ is a vector in G, and if we subtract it from the vector given in (8), we have a vector in G of the form

$$\sum_{j=1}^{k-1} r_j y^{(j)} + \nu_k^* \cdot c_k y^{(k)} \, ,$$

where $0 \le r_j < 1$, for $j = 1, \ldots, k-1$, and $0 < \nu_k^* < 1$. This contradicts the definition of c_k [see (5)]. Therefore our assumption that ν_k is not an integer must be false, and the $x^{(1)}, \ldots, x^{(k)}$ constructed above form a basis for G. This completes the proof of Theorem 18, and also of Theorem 17.

§3. Relation between different bases for a lattice

A lattice $\wedge$ of rank r can be obtained by starting with r arbitrary, linearly independent vectors $y^{(1)}, \ldots, y^{(r)}$ and considering all vectors of the form

$$g_1 y^{(1)} + \ldots + g_r y^{(r)} \, ,$$

where $g_1, \ldots, g_r$ are integers. By the construction in § 2, we may find r vectors $x^{(1)}, \ldots, x^{(r)}$ in $\wedge$, which will serve as a basis for $\wedge$.

Suppose the vectors $z^{(1)}, \ldots, z^{(r)}$ are also a basis for $\wedge$. This implies that

$$(9) \qquad x^{(j)} = \sum_{k=1}^{r} h_{jk} z^{(k)} \, , \qquad j = 1, \ldots, r \, ,$$

where the h_{jk} are all integers. But also, since $x^{(1)}, \ldots, x^{(r)}$ form a basis, we must have

$$(10) \qquad z^{(k)} = \sum_{l=1}^{r} g_{kl} x^{(l)} \, , \qquad k = 1, \ldots, r \, ,$$

where the g_{kl} are all integers. Substituting for $z^{(k)}$ in (9) from (10), we find that $x^{(j)}$ can be represented as a linear combination of $x^{(1)}, \ldots, x^{(r)}$. Since $x^{(1)}, \ldots, x^{(r)}$ are linearly independent, the matrix of the resulting transformation must be the identity matrix. That is to say, the product of the matrix (h_{jk}) by the matrix (g_{kl}) must be the identity matrix, so that (h_{jk}) is the inverse of the matrix (g_{kl}), hence the determinant of (h_{jk}) is the reciprocal of the determinant of (g_{kl}). However the determinants must be integers, since all elements h_{jk}, g_{kl} are integers. It follows that $\det(h_{jk}) = \det(g_{kl}) = \pm 1$.

A matrix all of whose elements are integers and whose determinant is ± 1 is called a *unimodular matrix*. We have just proved that if $x^{(1)}, \ldots, x^{(r)}$, and $z^{(1)}, \ldots, z^{(r)}$, are two bases for a lattice, they must be connected by a *unimodular transformation*. The converse is also true.

If $x^{(1)}, \ldots, x^{(r)}$ form a basis for a lattice Λ, and $z^{(1)}, \ldots, z^{(r)}$ are obtained from $x^{(1)}, \ldots, x^{(r)}$ by a unimodular transformation, then $z^{(1)}, \ldots, z^{(r)}$ form a basis for Λ. This follows easily from the fact that the inverse of a unimodular matrix is unimodular, so that the $x^{(j)}$, $1 \leq j \leq r$, can be expressed as a linear combination of the $z^{(k)}$, $1 \leq k \leq r$, with integral coefficients, hence $z^{(1)}, \ldots, z^{(r)}$ also will be a basis. This completes the proof of

Theorem 19. *Let $x^{(1)}, \ldots, x^{(r)}$ be a basis for a lattice Λ. A necessary and sufficient condition that r linearly independent vectors*

$$z^{(1)}, \ldots, z^{(r)}$$

also form a basis for Λ is that the $z^{(k)}$, $1 \leq k \leq r$, should be obtained from the $x^{(j)}$, $1 \leq j \leq r$, by a unimodular transformation.

§4. Sub-lattices

Let Λ be a lattice of rank r, and $\mathcal{M}$ a lattice of rank r contained in Λ. Then $\mathcal{M}$ is called a *sub-lattice* of Λ.

Suppose that $x^{(1)}, \ldots, x^{(r)}$ form a basis for Λ, and $y^{(1)}, \ldots, y^{(r)}$ a basis for $\mathcal{M}$. Then

$$y^{(j)} = \sum_{k=1}^{r} g_{jk} x^{(k)}, \quad j = 1, \ldots, r,$$

where the g_{jk} are all integers. Let $m = |\det(g_{jk})|$. Then $m > 0$, since $y^{(1)}, \ldots, y^{(r)}$ are linearly independent, and m must be an integer, since the g_{jk} are all integers. If $m = 1$, then by Theorem 19, $y^{(1)}, \ldots, y^{(r)}$ also form a basis for Λ, and so $\mathcal{M}$ and Λ coincide.

The number m is uniquely determined by $\mathcal{M}$ and Λ, because choosing any other basis for $\mathcal{M}$ or Λ is equivalent to multiplying the above matrix (g_{jk}) either on the left or on the right by a unimodular matrix, and the absolute value of the determinant of the resulting matrix will be the same as that of the matrix (g_{jk}), since a unimodular matrix has determinant ± 1. We call m the *index of $\mathcal{M}$ in Λ.* (Written $m = [\Lambda : \mathcal{M}]$.)

As an illustration, let Λ be the lattice with basis vectors $(1,0)$, $(\frac{1}{2}, \frac{1}{2})$, while $\mathcal{M}$ is the lattice with basis vectors $(1,0)$, $(0,1)$. It is easy to see that $\mathcal{M}$ is completely contained in Λ. The index of $\mathcal{M}$ in Λ equals 2.

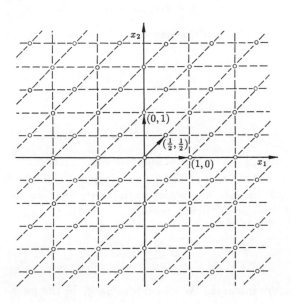

§5. Congruences relative to a sub-lattice

Let $\mathcal{M}$ be a sub-lattice of the lattice Λ, and let x, x' be two vectors belonging to Λ. We say that x *is congruent to* x' *modulo* $\mathcal{M}$, written $x \equiv x'(\bmod \mathcal{M})$ if the vector $x - x'$ belongs to $\mathcal{M}$. We say that two vectors in Λ which are congruent modulo $\mathcal{M}$ belong to the same *residue class modulo* $\mathcal{M}$. [Note that the relation '$\equiv$' is reflexive, symmetric and transitive.] We will prove

Theorem 20. *If $\mathcal{M}$ is a sub-lattice of the lattice Λ, then the number of different residue classes modulo $\mathcal{M}$ is m, the index of $\mathcal{M}$ in Λ.*

Suppose that $y^{(1)}, \ldots, y^{(r)}$ form a basis for $\mathcal{M}$. Then as in the proof of Theorem 18, we construct from them a basis for Λ. Let $x^{(1)}, \ldots, x^{(r)}$ be the basis for Λ, where by (5),

$$(11) \qquad x^{(k)} = c_{k1}y^{(1)} + \ldots + c_{k,k-1}y^{(k-1)} + c_k y^{(k)} , \qquad c_k > 0 ,$$

for $k = 1, \ldots, r$. Inverting these equations, we find that the $y^{(k)}$ are expressed in terms of the $x^{(k)}$ by a triangular matrix, as follows:

$$(12) \qquad y^{(k)} = \sum_{j=1}^{k} g_{kj} x^{(j)} , \qquad k = 1, \ldots, r ,$$

where (g_{kj}) is a triangular matrix, with $g_{kk} = 1/c_k > 0$, and all the g_{kj} are integers, since $x^{(1)}, \ldots, x^{(r)}$ form a basis for $\wedge$. Note that m, the index of M in $\wedge$, is the absolute value of the determinant of (g_{kj}), and therefore

$$(13) \qquad\qquad m = g_{11} g_{22} \ldots g_{rr} .$$

We now set up a representative system of residue classes modulo M. Consider the vectors

$$(14) \qquad\qquad h_1 x^{(1)} + \ldots + h_r x^{(r)} ,$$

where

$$(15) \qquad\qquad 0 \le h_j < g_{jj} , \quad j = 1, \ldots, r .$$

There are exactly $g_{11} g_{22} \ldots g_{rr} = m$ such vectors.

We must prove that any vector in $\wedge$ is congruent modulo M to a vector of the form (14), and that no two vectors of this form are congruent to each other modulo M. Let x be an arbitrary vector in $\wedge$, so that

$$x = \sum_{j=1}^{r} g_j x^{(j)} ,$$

where $g_1, \ldots, g_r$ are integers. We can represent g_r in the form

$$g_r = q_r g_{rr} + h_r ,$$

where q_r is an integer, and $0 \le h_r < g_{rr}$. Now using (12), we find that

$$x - q_r y^{(r)} = \sum_{j=1}^{r-1} g'_j x^{(j)} + h_r x^{(r)} ,$$

where $g'_1, \ldots, g'_{r-1}$ are integers. Proceeding in the same way, we can again reduce g'_{r-1}, and so on, until finally we have

$$x - \sum_{j=1}^{r} q_j y^{(j)} = \sum_{j=1}^{r} h_j x^{(j)} ,$$

or

$$x \equiv \sum_{j=1}^{r} h_j x^{(j)} \pmod{\mathsf{M}} .$$

Suppose two of the vectors defined by (14) lie in the same residue class modulo M, which means that their difference, say z, lies in M. Let

$$z = \sum_{j=1}^{r} n_j x^{(j)} .$$

Then we have

(16) $-g_{jj} < n_j < g_{jj}$, $j = 1, \ldots, r$.

And since $z \in \mathsf{M}$, we have also

$$z = \sum_{j=1}^{r} m_j y^{(j)} ,$$

where $m_1, \ldots, m_r$ are integers. Suppose $z \neq 0$, and let m_k denote the last coordinate which is not zero. Because of (12), we then have

$$|n_k| = |m_k g_{kk}| \geq g_{kk} .$$

This contradicts (16). Hence $m_1 = m_2 = \ldots = m_r = 0$, and z must be the zero vector.

§6. The number of sub-lattices with given index

Let m be a positive integer, and Λ a lattice of rank r. We will prove that there are at most m^{r^2} sub-lattices of index m in Λ.

As before let $x^{(1)}, \ldots, x^{(r)}$ form a basis for Λ, and $y^{(1)}, \ldots, y^{(r)}$ a basis for M, a sub-lattice of Λ, of index m in Λ. We have

$$y^{(k)} = \sum_{j=1}^{r} u_{kj} x^{(j)} , \quad k = 1, \ldots, r ,$$

where the u_{kj} are all integers, and $\det(u_{kj}) = \pm m$. Denote by U the matrix (u_{kj}). Let $\widetilde{\mathsf{M}}$ be another sub-lattice of index m in Λ, with $z^{(1)}, \ldots, z^{(r)}$ as a basis. We then have

$$z^{(k)} = \sum_{j=1}^{r} \widetilde{u}_{kj} x^{(j)} , \quad k = 1, \ldots, r ,$$

where the $\widetilde{u}_{kj}$ are all integers, and $\det(\widetilde{u}_{kj}) = \pm m$. Let $\widetilde{U} = (\widetilde{u}_{kj})$.

We first show that if $u_{kj} \equiv \widetilde{u}_{kj} \pmod{m}$ for all $k = 1, \ldots, r$ and $j = 1, \ldots, r$, then M is identical with $\widetilde{\mathsf{M}}$. This will follow from Theorem 19, if we show that the $y^{(k)}$ can all be obtained from the $z^{(k)}$ by a unimodular transformation. Now the matrix that transforms the $z^{(k)}$ into the $y^{(k)}$ is $U\widetilde{U}^{-1}$. The elements of $\widetilde{U}^{-1}$ are obtained by taking the cofactors of $\widetilde{U}$ and dividing by the determinant of $\widetilde{U}$, which is $\pm m$. [If $A = (a_{ij})$ is an $r \times r$ matrix, the determinant of the submatrix M_{ij} obtained from A by erasing the ith row and j^{th} column is called the *minor* corresponding to a_{ij}. The *cofactor* A_{ij} is defined to be $(-1)^{i+j} \det M_{ij}$. If A is non-singular, the (i, j) entry in A^{-1} is $\frac{A_{ji}}{\det A}$.] Therefore $m\widetilde{U}^{-1}$ is a matrix with integral elements. Since $U \equiv \widetilde{U} \pmod{m}$, we have $mU\widetilde{U}^{-1} \equiv m\widetilde{U}\widetilde{U}^{-1} \equiv mE \equiv 0 \pmod{m}$, where E is the $r \times r$ unit matrix.

This means that all elements of $mU\widetilde{U}^{-1}$ are divisible by m, and therefore all elements of $U\widetilde{U}^{-1}$ are integers. But $|\det(U\widetilde{U}^{-1})| = \frac{m}{m} = 1$. This proves that $U\widetilde{U}^{-1}$ is a unimodular matrix, and therefore Λ and $\widetilde{\Lambda}$ are identical.

This shows that there are at most m possible values for any element of U, such that the corresponding sub-lattices of Λ are different, and since there are r^2 elements in U, the total number of possibilities for U is m^{r^2}.

Lecture VI

The preceding lecture has shown that in any discrete vector group G of rank r, there exists a basis, that is r linearly independent vectors $x^{(1)}, \ldots, x^{(r)}$, such that any vector x belonging to G can be written as

$$x = g_1 x^{(1)} + \ldots + g_r x^{(r)} ,$$

where $g_1, \ldots, g_r$ are integers.

In this lecture we will analyze vector groups which are not discrete. We will show that the closure of any such group is the direct sum of a lattice and an s-dimensional linear manifold. This important result will then be applied to the discussion of the approximate solution of linear equations.

§1. Local rank of a vector group

Let G be a vector group in $\mathbb{R}^n$, of rank r. Let $\epsilon > 0$. Consider all the vectors $x = (x_1, \ldots, x_n)$ of length $|x| = (x_1^2 + \ldots + x_n^2)^{1/2} < \epsilon$. Suppose $r(\epsilon)$ is the maximum number of linearly independent vectors in G whose length is less than ϵ. It is clear that $0 \leq r(\epsilon) \leq r$. We can also see that $r(\epsilon)$ is a non-decreasing function of ϵ. As a bounded monotone function, $r(\epsilon)$ tends to a limit both as $\epsilon \to \infty$, and $\epsilon \to 0$. The limit as $\epsilon \to \infty$ is obviously r, the rank of the group. Let the limit of $r(\epsilon)$ as $\epsilon \to 0$ be denoted by s. We call s the *local rank* of the vector group G. Since $r(\epsilon)$ is an integer, and $0 \leq r(\epsilon) \leq r$, it follows that s is also an integer, with $0 \leq s \leq r$, and, moreover, for sufficiently small ϵ, we have

(1) $$r(\epsilon) = s .$$

§2. Decomposition of a general vector group

If $s = 0$, then G is a discrete vector group [and Lecture V has shown that it is a lattice].

Suppose that s is different from zero. Choose s linearly independent vectors $y^{(1)}, \ldots, y^{(s)}$ in G, such that their lengths are less than the ϵ for which (1) is satisfied. These s vectors span an s-dimensional linear manifold E, which consists of all vectors of the form

(2) $$\lambda_1 y^{(1)} + \ldots + \lambda_s y^{(s)} ,$$

where $\lambda_1, \ldots, \lambda_s$ are real numbers.

The linear manifold E is independent of the choice of the vectors $y^{(1)}, \ldots, y^{(s)}$. For suppose we had chosen the vectors $z^{(1)}, \ldots, z^{(s)}$ instead, which span a linear manifold E'. If $E' \neq E$, this would imply that at least one of the vectors $z^{(1)}, \ldots, z^{(s)}$ is linearly independent of $y^{(1)}, \ldots, y^{(s)}$. This contradicts our definition of s, because we now have $1 + s$ linearly independent vectors of length less than ϵ. Therefore $E' = E$.

Let D denote *the intersection of the vector group G with the linear manifold E*. D is not empty, since all the vectors $y^{(1)}, \ldots, y^{(s)}$ belong to it. Furthermore D is a subgroup of G, since if x and x' belong to D, then $x - x'$ belongs to G by the group property, and $x - x'$ belongs to E by linearity, so that $x - x'$ belongs to D.

We will now prove that the closure $\overline{D}$ of D is E. That is, given any element $\overline{y}$ in E, and an arbitrary real number $\delta > 0$, there exists an element y in D, such that

$$(3) \qquad\qquad |\overline{y} - y| < \delta .$$

Since $\overline{y}$ is in E, we have from (2)

$$(4) \qquad\qquad \overline{y} = \sum_{j=1}^{s} \lambda_j y^{(j)} ,$$

where $\lambda_j = g_j + r_j$, g_j is an integer, and

$$(5) \qquad\qquad 0 \leq r_j < 1 ,$$

for $j = 1, \ldots, s$. We may rewrite (4) as

$$\overline{y} = \sum_{j=1}^{s} g_j y^{(j)} + \sum_{j=1}^{s} r_j y^{(j)} = y + \sum_{j=1}^{s} r_j y^{(j)} ,$$

where y is an element of $G \cap E = D$. Now we have

$$|\overline{y} - y| \leq \sum_{j=1}^{s} \left| r_j y^{(j)} \right| < \sum_{j=1}^{s} \left| y^{(j)} \right| < s\epsilon ,$$

by using (5) and the definition of $y^{(j)}$. If we choose ϵ, such that $0 < \epsilon < \frac{\delta}{s}$, [and such that ϵ is sufficiently small for (1) to be satisfied, and possibly new vectors $y^{(1)}, \ldots, y^{(s)}$ of length less than ϵ, to enable this proof to go through], then $|\overline{y} - y| < \delta$, which proves (3).

If $s = r$, then D is the whole vector group G. Suppose now that $r - s = q$, a positive integer. This means that there exist, in G, q linearly independent vectors $x^{(1)}, \ldots, x^{(q)}$ which are linearly independent of the s vectors $y^{(1)}, \ldots, y^{(s)}$. It follows that any element x in G may be expressed uniquely as follows:

$$x = \sum_{j=1}^{s} \lambda_j y^{(j)} + \sum_{j=1}^{q} \mu_j x^{(j)} ,$$

where $\lambda_1, \ldots, \lambda_s$ and $\mu_1, \ldots, \mu_q$ are real numbers. If we set

$$\overline{y} = \sum_{j=1}^{s} \lambda_j y^{(j)} ,$$

then $\overline{y}$ lies in E. Since we do not yet know that $\overline{y}$ belongs to G, we cannot yet assert that $\overline{y}$ belongs to D. Now

(6)
$$x = \overline{y} + \sum_{j=1}^{q} \mu_j x^{(j)} , \quad \overline{y} \in E ,$$

which shows that any x in G can be decomposed uniquely into the sum of an element $\overline{y}$ from E, and a linear combination of the vectors $x^{(1)}, \ldots, x^{(q)}$. Note that the coordinates $\mu_1, \ldots, \mu_q$ of the element $x \in G$ with respect to the vectors $x^{(1)}, \ldots, x^{(q)}$ are uniquely determined by x, but two different elements in G may have the *same* μ-coordinates.

Consider the subset L of $\mathbb{R}^n$, consisting of all vectors a given by $a = \sum_{j=1}^{q} \mu_j x^{(j)}$ in the decomposition (6), where x runs through the elements of G. We shall prove that L is a lattice. First, we prove that it is a vector group, and then that this vector group is discrete.

Suppose a' corresponding to an element $x' \in G$, and a'' corresponding to an element $x'' \in G$, belong to L. This means that

$$x' = \overline{y}' + a' , \quad \overline{y}' \in E ,$$

and

$$x'' = \overline{y}'' + a'' , \quad \overline{y}'' \in E .$$

Subtracting, we have

$$x' - x'' = \overline{y}' - \overline{y}'' + a' - a'' .$$

Since $x' - x''$ belongs to G, and $\overline{y}' - \overline{y}''$ belongs to E, and $a' - a''$ is a linear combination of $x^{(1)}, \ldots, x^{(q)}$, this shows that $a' - a''$ belongs to L, so that L is a vector group.

If L were not discrete, then given any $\delta > 0$, there would exist a non-zero element a of L, such that $|a| < \delta$. Let x be a vector in G corresponding to a; that is to say, one for which

$$x = \overline{y} + a , \quad \overline{y} \in E .$$

We do not yet know that $\overline{y}$ is in D, but since $E = \overline{D}$, the closure of D, there exists in D a vector y, such that $|\overline{y} - y| < \delta$. Now

$$|x - y| \leq |x - \overline{y}| + |\overline{y} - y| < 2\delta ,$$

since $|x - \overline{y}| = |a| < \delta$. Given $\epsilon > 0$, we can choose δ so small that $|x - y| < \epsilon$. The vector $z = x - y$ is in G, but not in E (by the assumption that $a \neq 0$), and its length is less than ϵ. But if for ϵ sufficiently small, z is not the zero vector, then we would have besides $y^{(1)}, \ldots, y^{(s)}$ a vector z, which is linearly independent of them, and which has a length less than ϵ. This contradicts the definition of s. Hence we must have $a = 0$. It follows that L is discrete.

Since L is of rank q, it has a basis $z^{(1)}, \ldots, z^{(q)}$ of vectors in G. The decomposition in (6) then becomes

$$x = \overline{y} + \sum_{j=1}^{q} g_j z^{(j)} , \quad \overline{y} \in E ,$$

where $g_1, \ldots, g_q$ are integers. The sum $\sum_{j=1}^{q} g_j z^{(j)}$ obviously belongs to G, and since $x \in G$, it follows that $\overline{y} \in G$, and since $\overline{y} \in E$, we see that $\overline{y} \in D$. Thus we have

Theorem 21. *Let G be a vector group of rank r, and let s be its local rank. Let $\epsilon > 0$, such that $r(\epsilon) = s$ [see § 1]. Let $y^{(1)}, \ldots, y^{(s)}$ be s linearly independent vectors in G, of length less than ϵ. Let E be the linear manifold spanned by these s vectors, and let D be the intersection of G and E. Then G can be written as the direct sum of D and a lattice L whose rank is $r - s$.*

Now consider $\overline{G}$, the closure of $G \subset \mathbb{R}^n$. Then $\overline{G}$ is also a vector group. If $\overline{x}$ is a limit point of a sequence of vectors $x^{(1)}, x^{(2)}, \ldots$ in G, and $\overline{y}$ a limit point of a sequence of vectors $y^{(1)}, y^{(2)}, \ldots$ in G, then $\overline{x} - \overline{y}$ is a limit point of the sequence of vectors $x^{(1)} - y^{(1)}, x^{(2)} - y^{(2)}, \ldots$ also in G, hence $\overline{x} \in \overline{G}, \overline{y} \in \overline{G}$ implies that $\overline{x} - \overline{y} \in \overline{G}$. We will now prove

Theorem 22. *The closure $\overline{G}$ of any vector group G may be written as the direct sum of E and L, where E and L are as in Theorem 21.*

To the vectors defined previously, namely

$$y^{(1)}, \ldots, y^{(s)} , \quad z^{(1)}, \ldots, z^{(q)} ,$$

join $n - (q + s) = n - r$ linearly independent vectors

$$t^{(1)}, \ldots, t^{(n-r)} ,$$

so as completely to span $\mathbb{R}^n$.

If x is any vector in $\mathbb{R}^n$, then

$$x = \sum_{j=1}^{s} \lambda_j y^{(j)} + \sum_{j=1}^{q} \mu_j z^{(j)} + \sum_{j=1}^{n-r} \nu_j t^{(j)} ,$$

where $\lambda_1, \ldots, \lambda_s; \mu_1, \ldots, \mu_q; \nu_1, \ldots, \nu_{n-r}$ are all real numbers. Suppose that x is also in $\overline{G}$. Then there exists a sequence of elements in G, converging to x, that is to say

$$x = \lim_{l \to \infty} \left(\sum_{j=1}^{s} \lambda_{lj} y^{(j)} + \sum_{j=1}^{q} \mu_{lj} z^{(j)} \right) ,$$

where the coefficients μ_{lj}, $j = 1, \ldots, q$, are integers for all l, and the coefficients λ_{lj}, $j = 1, \ldots, s$ are certain real numbers for all l. This implies that $\nu_1 = \nu_2 = \ldots = \nu_{n-r} = 0$, and that $\mu_1, \ldots, \mu_q$ are integers. Hence

$$x = \bar{y} + \sum_{j=1}^{q} \mu_j z^{(j)} \; ,$$

where $\bar{y} \in E$, and $\mu_1, \ldots, \mu_q$ are integers, which proves that $\bar{G}$ is contained in the direct sum of E and L. But by Theorem 21 G *is* the direct sum of D and L, where D is the intersection of G and E, hence $\bar{G}$ contains the direct sum of $\bar{D} = E$ and L, so that Theorem 22 is proved.

Illustration. To illustrate these results we shall investigate all possible vector groups in two dimensions. Now $n = 2$, and the possible values of r are $0, 1, 2$.

If $r = 0$, then G consists solely of the zero vector.

If $r = 1$, then all the vectors in G lie on a straight line through the origin. If $s = 0$, the elements of G are integral multiples of a fixed vector. If $s = 1$, the elements of G are everywhere dense on this straight line, and $\bar{G}$, the closure of G, is the entire straight line.

If $r = 2$, and $s = 0$, then G is a lattice of rank 2. If $s = 1$, then E is a straight line through the origin, and L is a lattice of rank 1, and G is everywhere dense on parallel straight lines. If $s = 2$, then $\bar{G}$ is the whole plane.

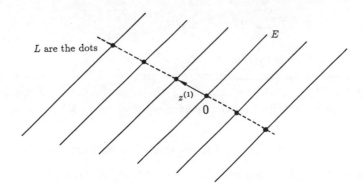

§3. Characters of vector groups

Let χ be a real-valued continuous function on $\mathbb{R}^n$, and suppose that it satisfies the functional equation

$$\chi(x + y) = \chi(x) + \chi(y) \; , \quad x, y \in \mathbb{R}^n \; .$$

Then it can be proved that the general solution of this equation which is continuous, is any homogeneous, linear function of the coordinates of x, that is

$$\chi(x) = x_1 \xi_1 + x_2 \xi_2 + \ldots + x_n \xi_n \; ,$$

where $x = (x_1, \ldots, x_n)$, [recall that we have chosen an orthonormal basis in $\mathbb{R}^n$, cf. Lecture I], and $\xi_1, \ldots, \xi_n$ are arbitrary real numbers. If we consider $\xi_1, \ldots, \xi_n$ as the coordinates of a vector ξ in $\mathbb{R}^n$, then we may write

$$\chi(x) = x \cdot \xi \, ,$$

where $x \cdot \xi$ is the scalar product of the vectors x and ξ. Note that the function χ is uniquely determined by the vector ξ (and vice versa).

Such a function χ is called a *character of the vector group G*, if χ is integral for all x in G. We know that there exist characters for every vector group G, since $\chi(x) \equiv 0$ is a character. We have already seen that every element x of G can be expressed as

$$x = \sum_{j=1}^{s} \lambda_j y^{(j)} + \sum_{j=1}^{q} \mu_j z^{(j)} \quad \text{[see § 2 for the notation]} \, ,$$

where $\lambda_1, \ldots, \lambda_s$ are certain real numbers, and $\mu_1, \ldots, \mu_q$ are integers. As before we may introduce $n - r$ linearly independent vectors

$$t^{(1)}, \ldots, t^{(n-r)} \, ,$$

which together with the vectors $y^{(1)}, \ldots, y^{(s)}$, $z^{(1)}, \ldots, z^{(q)}$ span $\mathbb{R}^n$, so that any vector x in $\mathbb{R}^n$ may be written

$$x = \sum_{j=1}^{s} \lambda_j y^{(j)} + \sum_{j=1}^{q} \mu_j z^{(j)} + \sum_{j=1}^{n-r} \nu_j t^{(j)} \, ,$$

were *now* the coefficients λ_j, μ_j, ν_j are *real* numbers.

[We change now the Euclidean structure of $\mathbb{R}^n$ in such a way that $y^{(1)}, \ldots, y^{(s)}$, $z^{(1)}, \ldots, z^{(q)}$, $t^{(1)}, \ldots, t^{(n-r)}$ become the orthonormal basis with respect to which we write the coordinates of a vector in $\mathbb{R}^n$.]

§4. Conditions on characters

Suppose that χ is a character of the vector group G, and that it is determined by the vector $\xi = (a_1, \ldots, a_s, b_1, \ldots, b_q, c_1, \ldots, c_{n-r})$, so that, with the above basis in $\mathbb{R}^n$, we have, for $x \in \mathbb{R}^n$, $x = (\lambda_1, \ldots, \lambda_s, \mu_1, \ldots, \mu_q, \nu_1, \ldots, \nu_{n-r})$,

$$\chi(x) = \sum_{j=1}^{s} \lambda_j a_j + \sum_{j=1}^{q} \mu_j b_j + \sum_{j=1}^{n-r} \nu_j c_j \, .$$

It is easy to see that χ *is* a character of G, *if*
   (i) $a_1 = a_2 = \ldots = a_s = 0$,
and
   (ii) $b_1, \ldots, b_q$ are integers (since, for $x \in G$, $\mu_1, \ldots, \mu_q$ are integers).
   Note that $c_1, \ldots, c_{n-r}$ can be arbitrary real numbers, since for $x \in G$, we have $\nu_1 = \nu_2 = \ldots = \nu_{n-r} = 0$.

These conditions are also necessary. Suppose condition (ii) is not satisfied, so that b_1, say, is not an integer. Then $\chi\left(z^{(1)}\right) = b_1$ is not an integer, and yet $z^{(1)}$ is an element of G. This shows that χ cannot be a character. Condition (i) is necessary, because for $x \in D$ we have $\mu_j = 0$ for $j = 1, \ldots, q$, so that $\chi(x) = \sum_{j=1}^{s} \lambda_j a_j$. Now the set D is dense in the linear manifold E, and χ is a continuous function taking on only integral values in this dense subset. This implies that χ is constant on E, for otherwise it would have a jump of at least one, at two neighbouring points, which contradicts the continuity of χ. Since $\chi(0) = 0$, the constant must be zero, so that $\chi(x) = 0$, for $x \in E$, which proves that $a_1 = a_2 = \ldots = a_s = 0$.

§5. Duality theorem for character groups

Let x be a vector in $\mathbb{R}^n$. How can we tell whether x belongs to $\overline{G}$, the closure of G, which is a vector group? Of course, we can construct E and L [see § 2], but a more useful method is given by

Theorem 23. *Let G be a vector group in $\mathbb{R}^n$. A vector x belongs to $\overline{G}$, if and only if $\chi(x)$ is an integer for all characters χ of G.*

The proof of the necessity part is trivial. If $x \in \overline{G}$, there exists a sequence $x^{(1)}, x^{(2)}, \ldots$ of elements of G converging to x. Since χ is continuous, $\chi\left(x^{(l)}\right)$ converges to $\chi(x)$ as $l \to \infty$. Since $\chi\left(x^{(l)}\right)$ is an integer, for all $l = 1, 2, \ldots$, it follows that $\chi(x)$ is an integer.

The sufficiency part of the proof depends upon the representation of x given before:

$$x = \sum_{j=1}^{s} \lambda_j y^{(j)} + \sum_{j=1}^{q} \mu_j z^{(j)} + \sum_{j=1}^{n-r} \nu_j t^{(j)} \ .$$

Suppose x is *not* in $\overline{G}$. Then either one of the coefficients $\nu_1, \ldots, \nu_{n-r}$ is *not* zero, or all of the ν's are zero but one of the numbers $\mu_1, \ldots, \mu_q$ is *not* an integer. We show that in these cases there exist characters χ for which $\chi(x)$ is not an integer.

If, for example, $\nu_1 \neq 0$, set $c_1 = \frac{1}{2\nu_1}$, $c_j = 0$ for $j = 2, \ldots, n - r$; $b_j = 0$ for $j = 1, \ldots, q$; and $a_j = 0$ for $j = 1, \ldots, s$. Then the vector $\xi = (a_1, \ldots, a_s, b_1, \ldots, b_q, c_1, \ldots, c_{n-r})$ determines a character $\chi(x) = x \cdot \xi$ [see § 4] for which we have

$$\chi(x) = \frac{\nu_1}{2\nu_1} = \frac{1}{2} \ ,$$

which is not an integer.

If $\nu_j = 0$ for $j = 1, \ldots, n - r$, but, for example, μ_1 is not an integer, set $b_1 = 1, b_j = 0$ for $j = 2, \ldots, q$; $c_j = 0$ for $j = 1, \ldots, n - r$; and $a_j = 0$ for $j = 1, \ldots, s$; then $\chi(x) = \mu_1$, which is not an integer, which completes the proof.

This theorem can be interpreted in another way. Represent the character χ by the vector ξ which determines it, and the character χ' by the corresponding vector ξ'. It is easy to see that these 'character vectors' form a vector group, because $x \cdot (\xi - \xi') = x \cdot \xi - x \cdot \xi'$ is an integer for all $x \in G$, so that $\xi - \xi'$ also represents a character of G.

The group of these vectors is called the *character group* of G. We denote it by $\widehat{G}$. Now $\widehat{G} \subset \mathbb{R}^n$ is closed, because if a sequence $\xi^{(l)}$, $l = 1, 2, \ldots$, from $\widehat{G}$, converges to ξ, as $l \to \infty$, then $x \cdot \xi^{(l)}$ converges to $x \cdot \xi$ as $l \to \infty$, so that if $x \cdot \xi^{(l)}$ is an integer for all l, then $x \cdot \xi$ is also an integer, and ξ represents a character, so belongs to $\widehat{G}$.

Now Theorem 23 can be restated as

Theorem 23'. *The character group $(\widehat{G})\widehat{}$ of the character group $\widehat{G}$ is the closure of the original vector group G; symbolically $(\widehat{G})\widehat{} = \overline{G}$.*

The proof is simple. We have defined $\widehat{G}$ as the set of all vectors ξ, such that $x \cdot \xi$ is an integer, for all $x \in G$. Let x' determine a character of $\widehat{G}$. That is to say, $\xi \cdot x'$ is an integer, where x' is fixed, and ξ runs over all elements of $\widehat{G}$. We know by Theorem 23 that all vectors x' belonging to $\overline{G}$ determine characters of $\widehat{G}$. Suppose x' is not in $\overline{G}$. Then again, by Theorem 23, there exists a vector $\xi \in \widehat{G}$, such that $\xi \cdot x'$ is not an integer, so that x' cannot be in the character group of $\widehat{G}$, which proves Theorem 23'.

This duality theorem is similar to the famous duality theorem of Pontrjagin for locally compact abelian groups, and we shall apply it to the discussion of a theorem of Kronecker.

§6. Kronecker's approximation theorem

Given a set of m vectors

$$a^{(1)}, a^{(2)}, \ldots, a^{(m)}$$

in $\mathbb{R}^n$, they generate a vector group G containing all vectors of the form $\sum_{j=1}^{m} g_j a^{(j)}$, where the g_j, $j = 1, \ldots, m$, are integers. Note that $a^{(1)}, \ldots, a^{(m)}$ need not be a basis for G, since they might not be linearly independent.

Given a vector b in $\mathbb{R}^n$, when does b belong to $\overline{G}$? Denote the components of $a^{(j)}$ by $a_1^{(j)}, \ldots, a_n^{(j)}$, for $j = 1, \ldots, m$. Then the question is: do there exist integers $g_1, \ldots, g_m$, such that

$$
\begin{aligned}
a_1^{(1)} g_1 + a_1^{(2)} g_2 + \ldots + a_1^{(m)} g_m &= b_1 + \epsilon_1 \, , \\
a_2^{(1)} g_1 + a_2^{(2)} g_2 + \ldots + a_2^{(m)} g_m &= b_2 + \epsilon_2 \, , \\
&\vdots \\
a_n^{(1)} g_1 + a_n^{(2)} g_2 + \ldots + a_n^{(m)} g_m &= b_n + \epsilon_n \, ,
\end{aligned}
$$

(7)

where $|\epsilon_j| < \epsilon$, $j = 1, \ldots, n$, for an arbitrarily given $\epsilon > 0$?

The similarity of this question with that which led to Theorem 23 is apparent. The answer is the same: let ξ be the vector defining a character of G. Since the vectors $a^{(1)}, \ldots, a^{(m)}$ are in G, we must have

$$(8) \qquad a^{(j)} \cdot \xi = \text{an integer} , \quad \text{for } j = 1, \ldots, m ;$$

but these conditions are sufficient, since the vectors $a^{(1)}, \ldots, a^{(m)}$ generate G. An application of Theorem 23 gives

Theorem 24. *Given $\epsilon > 0$, the system of equations (7) will have a solution in integers $g_1, \ldots, g_m$ if and only if*

$$(9) \qquad b \cdot \xi = \text{an integer} ,$$

for all ξ which determine a character of G; that is to say, for all ξ which satisfy condition (8).

Note the analogy of this theorem with the corresponding theorem about *real* solutions of linear equations. [Bôcher, *Introduction to Higher Algebra.*] When does the system in (7), with $\epsilon_1 = \epsilon_2 = \ldots = \epsilon_n = 0$ have real solutions? The answer is that it will have real solutions if and only if

$$b \cdot \xi = 0 ,$$

for all ξ satisfying the conditions

$$a^{(j)} \cdot \xi = 0 , \quad \text{for } j = 1, \ldots, m .$$

The content of Theorem 24 may appear puzzling at first. We might, for example, try to apply it as follows:

Let $m = n$ and suppose we have found integers $g_1, \ldots, g_n$ satisfying (7) for certain values $\epsilon_1, \ldots, \epsilon_n$, such that $|\epsilon_j| < \epsilon$, $j = 1, \ldots, n$, with $\epsilon > 0$. Suppose we now look for solutions of (7) within an accuracy ϵ', where $0 < \epsilon' < \epsilon$. We find new integers $g'_1, \ldots, g'_n$ satisfying (7) with $\epsilon'_1, \ldots, \epsilon'_n$ in place of $\epsilon_1, \ldots, \epsilon_n$, where $|\epsilon'_j| < \epsilon' < \epsilon$, for $j = 1, \ldots, n$. Now a slight change in ϵ appears to produce a change of at least one in the value of at least one of $g_1, \ldots, g_n$. This seems a contradiction to the known fact that the solution vector of a system of the form (7), of n equations in n unknowns, is a continuous function of the coordinates of the vector on the right-hand side.

The contradiction is removed, when we realize (see the example below) that in this case $(m = n)$ a solution of (7) in integers exists only if there exists a solution of (7) in integers *with $\epsilon_1 = \epsilon_2 = \ldots = \epsilon_n = 0$.*

As an illustration, let $n = 2$, $a^{(1)} = (\sqrt{2}, 1)$, $a^{(2)} = (0, \sqrt{3})$, and consider the following set of equations:

$$(10) \qquad \begin{aligned} \sqrt{2} \cdot g_1 + 0 \cdot g_2 &= b_1 + \epsilon_1 , \\ 1 \cdot g_1 + \sqrt{3} \cdot g_2 &= b_2 + \epsilon_2 . \end{aligned}$$

First we determine the vectors $\xi = (\xi_1, \xi_2)$ which represent the characters of the vector group generated by $a^{(1)}$ and $a^{(2)}$. We must have

$$\sqrt{2} \cdot \xi_1 + 1 \cdot \xi_2 = \text{an integer} = \mu_1 , \quad \text{say} ,$$

$$0 \cdot \xi_1 + \sqrt{3} \cdot \xi_2 = \text{an integer} = \mu_2 , \quad \text{say} .$$

From the second equation we have $\xi_2 = \mu_2/\sqrt{3}$; substituting this in the first equation, we get $\xi_1 = (\mu_1 - \mu_2/\sqrt{3})/\sqrt{2}$. In order that system (10) have an integral solution g_1, g_2, by Theorem 24 b_1 and b_2 must satisfy the relation

$$\frac{b_1}{\sqrt{2}} \left(\mu_1 - \frac{\mu_2}{\sqrt{3}} \right) + \frac{b_2}{\sqrt{3}} \cdot \mu_2 = \text{an integer} ,$$

or

$$\mu_1 \frac{b_1}{\sqrt{2}} + \mu_2 \left(\frac{b_2}{\sqrt{3}} - \frac{b_1}{\sqrt{6}} \right) = \text{an integer} ,$$

for *all* integral μ_1 and μ_2. This can happen only if

$$\frac{b_1}{\sqrt{2}} = \text{an integer} = g_1 , \quad \text{say} ,$$

and

$$\frac{b_2}{\sqrt{3}} - \frac{b_1}{\sqrt{6}} = \text{an integer} = g_2 , \quad \text{say} ,$$

so that we have

(11)
$$b_1 = \sqrt{2} \cdot g_1 ,$$
$$b_2 = g_1 + \sqrt{3} \cdot g_2 .$$

This shows that we can solve (10) in integers only when b_1, b_2 are of the form given in (11), and then it is possible to have $\epsilon_1 = \epsilon_2 = 0$.

If, however, we take $m > n$, we get a non-trivial result. Let $m = n+1$, and let

$$a^{(1)}, a^{(2)}, \ldots, a^{(n)}, a^{(n+1)}$$

be the column vectors of the matrix

$$\begin{pmatrix} -1 & 0 & \ldots & 0 & \lambda_1 \\ 0 & -1 & \ldots & 0 & \lambda_2 \\ \vdots & \vdots & \ddots & \vdots & \vdots \\ 0 & 0 & \ldots & -1 & \lambda_n \end{pmatrix} .$$

The system of equations in (7) then reduces to

(12)
$$\begin{aligned} \lambda_1 x &= b_1 &+& x_1 &+& c_1 , \\ \lambda_2 x &= b_2 &+& x_2 &+& \epsilon_2 , \\ &\vdots \\ \lambda_n x &= b_n &+& x_n &+& \epsilon_n , \end{aligned}$$

and an application of Theorem 24 gives

Theorem 25. *Given an arbitrary $\epsilon > 0$, there exist numbers $\epsilon_1, \ldots, \epsilon_n$, with $|\epsilon_j| < \epsilon$, for $j = 1, \ldots, n$, and integers $x_1, \ldots, x_n, x$, satisfying the equations given in (12), if and only if*

$$b_1 \xi_1 + \ldots + b_n \xi_n$$

is an integer whenever $\xi_1, \ldots, \xi_n$ and $\lambda_1 \xi_1 + \ldots + \lambda_n \xi_n$ are integers.

Suppose $\lambda_1, \ldots, \lambda_n$ are such that the equation

$$\lambda_1 g_1 + \lambda_2 g_2 + \ldots + \lambda_n g_n = g_{n+1} \, ,$$

where $g_1, \ldots, g_{n+1}$ are integers, implies that $g_1 = g_2 = \ldots = g_{n+1} = 0$ (which is the case if $\lambda_1, \ldots, \lambda_n, 1$ are linearly independent over $\mathbb{Q}$). Then the conditions of Theorem 25 are trivially satisfied.

As an illustration, let $\lambda_1 = \sqrt{2}$, $\lambda_2 = \sqrt{3}$. The numbers $1, \sqrt{2}, \sqrt{3}$ are linearly independent over the rationals. Therefore for any values of b_1, b_2 there exist integers g_1, g_2, g_3, such that

$$\sqrt{2} \cdot g_3 = b_1 + g_1 + \epsilon_1 \, ,$$

$$\sqrt{3} \cdot g_3 = b_2 + g_2 + \epsilon_2 \, ,$$

where $|\epsilon_j| < \epsilon$, $j = 1, 2$, for an arbitrary $\epsilon > 0$. We may state this in a more pictorial way as follows:

Consider in two-dimensional Euclidean space the points whose coordinates are $(g_3\sqrt{2}, g_3\sqrt{3})$, where g_3 is an integer. If we apply an integral translation to these points so that they land in the unit square, we will find that the translated points are dense in the unit square. One sees that if on a unit square billiard table we shoot a ball in the direction of the line whose slope is $\frac{\sqrt{2}}{\sqrt{3}}$, the ball will come arbitrarily close to every point on the table. [See E. Hlawka, *Theorie der Gleichverteilung*, p. 13.]

Lecture VII

§1. Periods of real functions

We shall use our results about the decomposition of vector groups to discuss the possible number of periods of real and complex functions.

Let f be a real-valued, continuous function defined on $\mathbb{R}^n$. A vector $p \in \mathbb{R}^n$ is called a *period* of f, if

$$f(x + p) = f(x), \quad \text{for all } x \in \mathbb{R}^n .$$

In particular, the zero vector is a period of f; and the difference of any two periods of f is a period of f. The set of all periods of f is a vector group, which is *closed* in $\mathbb{R}^n$, for let $p^{(1)}, p^{(2)}, \dots$ be a sequence of periods of f converging to some vector p. Then, by definition, $f(x + p^{(j)}) = f(x)$, and since f is continuous, we have $f(x + p) = f(x)$, so that p is also a period of f.

Suppose that a basis in $\mathbb{R}^n$ has been chosen. If $f(x)$ does not depend on the first component x_1 of the vector x, then any vector whose last $n - 1$ components are zero will be a period of f. We would like to eliminate such a trivial possibility. The situation is made more complicated because it is quite possible for $f(x)$ to depend on all the n components of x, but in such a way that a transformation of coordinates will reduce f to a function of less than n variables. For example, we might have

$$(1) \qquad\qquad f(x) = \tilde{f}(x_1 + x_2, x_3, \dots, x_n) .$$

In that case, any vector whose first component is the negative of the second component, and whose last $n - 2$ components are zero, will be a period of f. We will show how to exclude such a possibility.

Let C be a real, non-singular $n \times n$ matrix. We write

$$(2) \qquad\qquad g(y) = f(Cy), \quad y \in \mathbb{R}^n .$$

If p is a period of f, and if $p = Cq$, then q is a period of g, for

$$(3) \qquad g(y + q) = f(Cy + Cq) = f(Cy + p) = f(Cy) = g(y) .$$

We say that f is a *proper function*, if there exists no such matrix C for which $g(y)$ does not depend on the first component y_1 of y. It is very easy to prove

Theorem 26. *The periods of a proper function form a lattice.*

We have already seen that the periods of f form a closed vector group. Suppose the vector group is not discrete. Then, by Theorem 22, the group contains a linear manifold. This linear manifold must contain some straight line through the origin; that is to say, there exists a vector $p \neq 0$, such that λp is a period of f for all real values of λ, and so we have

$$(4) \qquad f(x + \lambda p) = f(x) , \quad \text{for all } x \in \mathbb{R}^n, \text{ and all real numbers } \lambda .$$

Introduce new coordinates, such that p is the new unit vector $(1, 0, \ldots, 0)$. Then for the transformed function g, we have, by (4),

$$(5) \qquad g(y_1 + \lambda, y_2, \ldots, y_n) = g(y_1, \ldots, y_n) ,$$

for all real numbers λ. If we put $\lambda = -y_1$, we see that $g(y_1, \ldots, y_n)$ does not depend on y_1, so that f could not be proper. This contradiction shows that the group of periods must be discrete, and therefore must form a lattice.

This implies that a proper function f has at most n linearly independent (over $\mathbb{Q}$) periods.

§2. Periods of analytic functions

A complex vector z is a vector with n complex components $z_1, \ldots, z_n$. The space of all such vectors will be denoted by $\mathbb{C}^n$. We define

$$|z| = (|z_1|^2 + |z_2|^2 + \ldots + |z_n|^2)^{1/2} .$$

Let f be a single-valued analytic function of the complex vector z. This means that if we start with any point in $\mathbb{C}^n$ where f is regular and continue the function along any closed regular path until we return to the original point, we will finish with the same value of the function as we started with.

The complex vector $p \in \mathbb{C}^n$ is a *period* of f, if

$$(6) \qquad f(z + p) = f(z) , \quad \text{for all } z \in \mathbb{C}^n .$$

This means that if formula (6) holds for all points z in a neighbourhood of any point z_0 at which f is regular, then it holds for all z in $\mathbb{C}^n$.

It is clear that the periods of f form a vector group. As before we can show that this vector group is closed. Let the sequence of periods $p^{(1)}, p^{(2)}, \ldots$ of f converge to a vector p. Let z_0 be a regular point of f. Then there exists a neighbourhood of z_0 in which f is regular, say $|z - z_0| < \epsilon$, for some $\epsilon > 0$. Then there exists an integer N, such that $|p - p^{(j)}| < \epsilon/2$, for $j \geq N$. Now, since $p^{(j)}$ is a period of f, we have

$$(7) \qquad f\left(z + p^{(j)}\right) = f(z) , \quad \text{for } |z - z_0| < \epsilon ,$$

(for a suitably chosen ϵ, perhaps smaller than the original one) and we have also that $z_0 + p^{(j)}$ is a regular point, since $f(z_0) = f\left(z_0 + p^{(j)}\right)$. Hence

f is regular in a sufficiently small neighbourhood U of $z_0 + p^{(j)}$, where $U = \left\{ z \in \mathbb{C}^n \,\middle|\, |z - (z_0 + p^{(j)})| < \epsilon \right\}$. For $|z - z_0| < \frac{\epsilon}{2}$, $|p - p^{(j)}| < \frac{\epsilon}{2}$ we have $z + p = z_0 + p^{(j)} + (z - z_0) + (p - p^{(j)}) \in U$, hence $f(z + p) = f(z)$ for all $z \in \mathbb{C}^n$.

Our previous results for real functions with real periods cannot be applied here. However the method of proof will be similar to that used before.

A function f of the above type is called a *proper function*, if there does not exist a complex linear transformation which eliminates any of the variables. We can prove

Theorem 27. *If f is a proper, single-valued analytic function of n complex variables, it has a most $2n$ linearly independent periods.*

Let $p = a + b\sqrt{-1}$ be a period of f, where a and b are real vectors, with components $a_1, a_2, \ldots, a_n$ and $b_1, b_2, \ldots, b_n$ respectively. Consider the vector $p^* \in \mathbb{R}^{2n}$, whose components are $a_1, a_2, \ldots a_n, b_1, b_2, \ldots, b_n$. We call p^* the image of p. It is clear that if p^* is the image of p, and q^* the image of q, then $p^* \pm q^*$ is the image of $p \pm q$. This fact enables us to consider the real vector group of images, instead of the original vector group of complex periods.

Suppose the vector group of images is not discrete. Then it contains a whole line passing through the origin; that is to say, there is a vector $p^* \neq 0$, such that λp^* belongs to the vector group, for every real λ. This implies that λp is a period of f for every real λ, hence

$$(8) \qquad f(z + \lambda p) = f(z) , \quad \text{for all } z \in \mathbb{C}^n, \text{ and all real numbers } \lambda .$$

Since f is analytic, the left-hand side of (8) is an analytic function of λ. Since (8) holds for all real λ, it must (by analytic continuation) hold also for all complex values of λ. Then, just as before, we can show that f is not a proper function. This contradiction shows that the image vector group, and also the original group of complex periods, must be discrete, and therefore it is a lattice.

The rank r of the lattice of images is not greater than $2n$. Therefore in the lattice of complex periods of f there exists a basis of r vectors $p^{(1)}, \ldots, p^{(r)}$, such that every period of f is an integral linear combination of these r periods. Since $r \leq 2n$, we have proved Theorem 27. Any such basis for the lattice of complex periods of f is called a *fundamental system*.

§3. Periods of entire functions

In case $n = 1$, we know (by Liouville's Theorem) that a proper entire function cannot have two linearly independent periods. This suggests

Theorem 28. *If f is a proper entire function of n variables, then it cannot have more than n linearly independent periods.* [An entire function of n variables is a function which can be represented by a power series which converges in the entire space $\mathbb{C}^n$.]

Since the vector group of periods is a lattice, all that has to be proved is that the lattice is of rank at most n. We have so far used the words "linearly independent" in the sense of "linearly independent over the reals" or "linearly independent over the rationals". We now introduce the notion of linear independence in the complex sense. We say that the vectors $p^{(1)}, \ldots, p^{(k)}$ in $\mathbb{C}^n$ are linearly independent over $\mathbb{C}$, if the equation

$$(9) \qquad \lambda_1 p^{(1)} + \lambda_2 p^{(2)} + \ldots + \lambda_k p^{(k)} = 0 ,$$

where $\lambda_1, \lambda_2, \ldots, \lambda_k$ are complex numbers, implies that

$$(10) \qquad \lambda_1 = \lambda_2 = \ldots = \lambda_k = 0 .$$

Proof of Theorem 28. Suppose now that there exist r (but not $r+1$) linearly independent periods in the complex sense. Denote them by $p^{(1)}, \ldots, p^{(r)}$. It is clear that r is equal to or less than n, since the periods are vectors in $\mathbb{C}^n$. With a suitable transformation of the coordinates it is possible to bring these r vectors into the r unit vectors: $e^{(1)}, \ldots, e^{(r)}$, where $e^{(k)}$ is the vector all of whose components are zero except the kth component which is 1. These unit vectors will be periods in the new coordinates, so that we have

$$(11) \qquad \tilde{f}\left(z + e^{(k)}\right) = \tilde{f}(z) , \quad k = 1, \ldots, r ; \quad \text{for all } z \in \mathbb{C}^n ,$$

where $\tilde{f}$ is the proper entire function into which f is transformed by the coordinate transformation just made.

Consider any *other* period p of $\tilde{f}$. Since there were only r linearly independent periods in the complex sense, we can write

$$(12) \qquad p = \sum_{j=1}^{r} p_j e^{(j)} ,$$

where $p_1, \ldots, p_r$ are complex numbers. We will show that they must be *real*. This implies that the lattice of periods of f in this coordinate system is a real lattice, and contains only r linearly independent vectors over the reals. From Lecture V we know that every lattice of rank r contains r linearly independent vectors which form a basis. Since $r \leq n$, this would prove Theorem 28.

In order to prove that $p_1, \ldots, p_r$ must be real, let $p_j = a_j + b_j \sqrt{-1}$, with a_j, b_j real, and let

$$a = (a_1, a_2, \ldots, a_r, 0, 0, \ldots, 0) ,$$
$$b = (b_1, b_2, \ldots, b_r, 0, 0, \ldots, 0) ,$$

be two vectors in $\mathbb{R}^n$, whose last $n - r$ components are zero. For a fixed vector $z \in \mathbb{C}^n$, consider $\tilde{f}(z + \lambda b)$ as a function of the complex variable λ. Let

$$(13) \qquad \lambda = \mu + \nu\sqrt{-1} , \quad \text{where } \mu \text{ and } \nu \text{ are real} .$$

Suppose now that b is not the zero vector. We write

$$(14) \qquad \lambda b = \mu b + \nu \left(-a + a + b\sqrt{-1}\right) = \mu b - \nu a + \nu p \ .$$

Here $\mu b - \nu a$ is a real vector, whose last $n - r$ components are zero. By subtracting from each component the largest integer (contained in it), we can write

$$(15) \qquad \mu b - \nu a = g + k \ , \quad \text{say} \ ,$$

where g is a vector with integral components and k is a vector all of whose components are greater than or equal to zero and less than 1. In the same way, if γ is the largest integer contained in ν, we can write

$$(16) \qquad \nu p = \gamma p + \kappa p \ ,$$

where γ is an integer, and $0 \leq \kappa < 1$. Combining (14) and (15) and (16), we get

$$(17) \qquad \lambda b = g + k + \gamma p + \kappa p = k + \kappa p + g + \gamma p \ ,$$

and we may write

$$(18) \qquad \widetilde{f}(z + \lambda b) = \widetilde{f}(z + k + \kappa p + g + \gamma p) \ .$$

Now since g is an integral vector whose last $n-r$ components are zero, it can be expressed as a linear combination of $e^{(1)}, \ldots, e^{(r)}$, with integral coefficients; since $e^{(1)}, \ldots, e^{(r)}$ are periods of $\widetilde{f}$, it follows that g is a period of $\widetilde{f}$. Also since γ is an integer, and p is a period of $\widetilde{f}$, γp is a period of $\widetilde{f}$. Equation (18) then simplifies to

$$(19) \qquad \widetilde{f}(z + \lambda b) = \widetilde{f}(z + k + \kappa p) \ .$$

Since f is assumed to be entire, $\widetilde{f}(z + \lambda b)$ is an entire function of λ. The vector $k + \kappa p$ lies in a bounded domain which does not depend on λ. Hence $\widetilde{f}(z + \lambda b)$, considered as a function of λ, is entire, and bounded for all $\lambda \in \mathbb{C}$. This implies that $\widetilde{f}(z + \lambda b)$ is a constant, independent of λ, but in that case, just as before, $\widetilde{f}$ cannot be a proper function. This contradiction shows that $b = 0$, so that all the periods of f are real in the new coordinate system. We saw before how from this fact we can prove Theorem 28.

We can easily construct a proper entire function of n variables, which has n linearly independent periods. For example, take

$$f(z) = e^{2\pi i z_1} + e^{2\pi i z_2} + \ldots + e^{2\pi i z_n} \ , \quad z \in \mathbb{C}^n \ , \quad z = (z_1, \ldots, z_n) \ ,$$

which has the periods $e^{(1)}, \ldots, e^{(n)}$, where $e^{(k)}$ is the unit vector defined previously.

§4. Minkowski's theorem on linear forms

As a simple consequence of a previous theorem [Lecture II, Theorem 13], we shall prove

Theorem 29. *Let $y_1, \ldots, y_n$ be n real linear forms given by*

$$
\begin{aligned}
y_1 &= a_{11}x_1 + \ldots + a_{1n}x_n, \\
&\vdots \\
y_n &= a_{n1}x_1 + \ldots + a_{nn}x_n.
\end{aligned}
$$
(20)

Let D be the absolute value of the determinant of the matrix (a_{jk}), $j, k = 1, \ldots, n$. If $D \neq 0$, there exist integers $x_1, \ldots, x_n$, not all zero, such that

$$
|y_1| \leq D^{1/n}, \quad \ldots, \quad |y_n| \leq D^{1/n}.
$$
(21)

This theorem can be interpreted in the following way: as $x_1, \ldots, x_n$ go over all integral values, $y = (y_1, \ldots, y_n)$ goes over all points of the lattice $\wedge$ in $\mathbb{R}^n$ determined by the vectors

$$
\begin{pmatrix} a_{11} \\ \vdots \\ a_{n1} \end{pmatrix}, \quad \ldots, \quad \begin{pmatrix} a_{1n} \\ \vdots \\ a_{nn} \end{pmatrix}.
$$

The volume of the parallelepiped formed by these vectors is D. Theorem 29 states that there exists a lattice point, distinct from the origin, such that all of its coordinates are equal to or less than $D^{1/n}$ in absolute value.

We can reduce this theorem to a consequence of Minkowski's First Theorem, by introducing the gauge function

$$
f(x) = \max\{|x_1|, \ldots, |x_n|\}, \quad x = (x_1, \ldots, x_n).
$$

It is obvious that f is an even, non-negative, positive-homogeneous function of degree 1 which vanishes only at $x = 0$. To check that f is a gauge function, we must prove that f satisfies the triangle inequality, that is

$$
f(x + w) \leq f(x) + f(w), \quad x, w \in \mathbb{R}^n.
$$

If $w = (w_1, \ldots, w_n)$, we have

$$
|x_j + w_j| \leq |x_j| + |w_j| \leq \overline{x} + \overline{w}, \quad 1 \leq j \leq n,
$$

where $\overline{x} = \max\{|x_1|, \ldots, |x_n|\}$, $\overline{w} = \max\{|w_1|, \ldots, |w_n|\}$. Therefore

$$
f(x + w) = \max_{1 \leq j \leq n} |x_j + w_j| \leq \overline{x} + \overline{w} = f(x) + f(w).
$$

Let $\mu = \min f(y)$, where y runs over the points of $\wedge$, different from the origin.

Minkowski's First Theorem states that

(22) $$V\mu^n \leq 2^n D ,$$

where V denotes the volume of the convex body $\{w \in \mathbb{R}^n | f(w) < 1\}$, which is an n-dimensional cube, with sides of length 2, so that actually $V = 2^n$. Hence

(23) $$\mu \leq D^{1/n} .$$

Let $y^* \neq 0$, $y^* = (y_1^*, \ldots, y_n^*)$, be a vector in $\wedge$ at which f attains its minimum μ. Then

$$\mu = f(y^*) = \max\{|y_1^*|, \ldots, |y_n^*|\} .$$

This implies, on using (23), that there exist integers $x_1^*, \ldots, x_n^*$, not all zero, such that

$$|y_1^*| \leq D^{1/n} , \quad \ldots, \quad |y_n^*| \leq D^{1/n} ,$$

which proves Theorem 29.

Suppose that $D = 1$. Theorem 29 states that there exists in $\wedge$ a vector, different from the origin, all of whose components are less than or equal to 1 in absolute value. We are going to investigate in which cases we can be sure that strict inequality holds.

We can give a sufficient condition very easily. Suppose the lattice $\wedge$ is generated by the following vectors:

(24)
$$\begin{pmatrix} 1 \\ 0 \\ 0 \\ 0 \\ \vdots \\ 0 \end{pmatrix}, \begin{pmatrix} a_{12} \\ 1 \\ 0 \\ 0 \\ \vdots \\ 0 \end{pmatrix}, \begin{pmatrix} a_{13} \\ a_{23} \\ 1 \\ 0 \\ \vdots \\ 0 \end{pmatrix}, \ldots, \begin{pmatrix} a_{1,n-1} \\ a_{2,n-1} \\ a_{3,n-1} \\ \vdots \\ 1 \\ 0 \end{pmatrix}, \begin{pmatrix} a_{1,n} \\ a_{2,n} \\ a_{3,n} \\ \vdots \\ a_{n-1,n} \\ 1 \end{pmatrix} .$$

It is obvious in this case that $D = 1$. If there exists a lattice-point in $\wedge$ such that all its coordinates are *less* than 1 in absolute value, we will show that this point must be the origin. For the general point of the lattice $\wedge$ has coordinates

(25)
$$\begin{array}{cccccccc}
g_1 & + & a_{12}g_2 & + & a_{13}g_3 & + \cdots + & a_{1,n-1}g_{n-1} & + & a_{1,n}g_n \\
 & & g_2 & + & a_{23}g_3 & + \cdots + & a_{2,n-1}g_{n-1} & + & a_{2,n}g_n \\
 & & & & g_3 & + \cdots + & a_{3,n-1}g_{n-1} & + & a_{3,n}g_n \\
 & & & & & & & \vdots & \vdots \\
 & & & & & & g_{n-1} & + & a_{n-1,n}g_n \\
 & & & & & & & & g_n ,
\end{array}$$

where $g_1, \ldots, g_n$ are integers. If these coordinates are all less than 1 in absolute value, we get

(26) $$|g_n| < 1 ,$$

(27) $$|g_{n-1} + a_{n-1,n}g_n| < 1 ,$$

and so on. Since g_n is an integer, (26) implies that $g_n = 0$. Using this in (27), and noting that g_{n-1} is an integer, we get $g_{n-1} = 0$. Continuing in the same way, we can show that $g_1, \ldots, g_n$ are all zero.

Instead of using the vectors in (24) as a basis for Λ, we may use any other basis for Λ. Since such a basis can be obtained from (24) by a unimodular transformation (given by an integral matrix with determinant ± 1), we may conjecture the following

Theorem 30. *A lattice with determinant* 1 *has no point, different from the origin, in the interior of the cube with centre at the origin and sides parallel to the coordinate axes and of length* 2, *if and only if there exists a basis for the lattice which can be expressed in the form* (24).

The necessity part is very difficult. Minkowski stated the theorem and proved it for $n = 2$ and 3. We shall prove these cases. The general case was proved in 1941 by Hajós.

Lecture VIII

§1. Completing a given set of vectors to form a basis for a lattice

Let $\wedge$ be a lattice of rank r in $\mathbb{R}^n$. Let E be a linear manifold through the origin in $\mathbb{R}^n$, of dimension q, $q < r$. Assume that we have q linearly independent vectors $x^{(1)}, \ldots, x^{(q)}$ in E, such that they form a basis for the lattice $\wedge \cap E$. Such a set of vectors will be called a set of *primitive vectors*. We shall show that there exist $r - q$ vectors $y^{(q+1)}, \ldots, y^{(r)}$, such that

$$(1) \qquad\qquad x^{(1)}, \ldots, x^{(q)} \ , \ y^{(q+1)}, \ldots, y^{(r)}$$

form a basis for $\wedge$.

The proof follows easily from Theorem 18 of Lecture V. Choose any $r - q$ vectors $x^{(q+1)}, \ldots, x^{(r)}$ in $\wedge$ which are linearly independent of each other, *and* also linearly independent of the given vectors $x^{(1)}, \ldots, x^{(q)}$. From these r linearly independent vectors $x^{(1)}, \ldots, x^{(r)}$ we can construct [Lecture V, Theorem 18] a basis $y^{(1)}, \ldots, y^{(r)}$ for $\wedge$, such that

$$
\begin{aligned}
y^{(1)} &= c_1 x^{(1)} \ , \\
y^{(2)} &= c_{21} x^{(1)} + c_2 x^{(2)} \ , \\
&\ \ \vdots \\
y^{(r)} &= c_{r1} x^{(1)} + c_{r2} x^{(2)} + \ldots + c_{r,r-1} x^{(r-1)} + c_r x^{(r)} \ .
\end{aligned}
$$

(2)

Solve the first q equations in (2) for $y^{(1)}, \ldots, y^{(q)}$. They belong to $\wedge$ and E, since they depend linearly only on $x^{(1)}, \ldots, x^{(q)}$, which, by hypothesis, belong to $\wedge \cap E$. Since $y^{(1)}, \ldots, y^{(r)}$ form a basis for the whole lattice $\wedge$, $y^{(1)}, \ldots, y^{(q)}$ form a basis for the lattice $\wedge \cap E$. We are given, however, that $x^{(1)}, \ldots, x^{(q)}$ form a basis for this lattice. The two bases must be connected by a unimodular transformation. This implies that c_j and c_{jk}, for $j = 1, \ldots, q$; $k = 1, \ldots, j - 1$ are integers. Therefore $y^{(1)}, \ldots, y^{(q)}$ can be expressed as linear combinations, with integer coefficients, of $x^{(1)}, \ldots, x^{(q)}$. This proves that

$$(3) \qquad\qquad x^{(1)}, \ldots, x^{(q)} \ , \ y^{(q+1)}, \ldots, y^{(r)}$$

is a basis for $\wedge$, and completes the proof of

Theorem 31. *Let $x^{(1)}, \ldots, x^{(q)}$ be q vectors which form a basis for the lattice of all vectors in the intersection of the lattice $\wedge$ of rank r and a linear manifold E through the origin in $\mathbb{R}^n$ of dimension q, $q < r$. Then there exist $r - q$ vectors $y^{(q+1)}, \ldots, y^{(r)}$, such that $x^{(1)}, \ldots, x^{(q)}, y^{(q+1)}, \ldots, y^{(r)}$ form a basis for $\wedge$.*

Consider the special case $q = 1$. E is now a straight line through the origin. The assumption is that there is a vector $x^{(1)} \neq 0$, such that (i) all points of the lattice $\wedge$ which belong to E are integral multiples of $x^{(1)}$. Another way of expressing this fact is that (ii) $x^{(1)}$ is a lattice point in E, different from the origin, which is nearest to the origin.

Theorem 31 now tells us that if $x^{(1)}$ satisfies any of the above two conditions, there exist $r - 1$ vectors which, together with $x^{(1)}$, form a basis for $\wedge$.

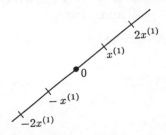

§2. Completing a matrix to a unimodular matrix

With the help of Theorem 31 we can now answer the following question:

Given n *integers* $g_1, \ldots, g_n$ as the first column of a matrix, when can we complete the matrix so that the result will be a unimodular matrix U?

It is clear that a necessary condition is that the greatest common divisor of $g_1, \ldots, g_n$ equals 1, for otherwise the determinant of U cannot be ± 1, since it would have a common divisor as a factor. We shall show that this condition is also sufficient.

We first change the problem so that Theorem 31 will be directly applicable. Let $u^{(j)}$ be the jth column vector of U, $u^{(1)}$ having as its components $g_1, \ldots, g_n$. We may now consider $u^{(1)}, \ldots, u^{(n)}$ as a basis for the lattice of all g-points, since $u^{(1)}, \ldots, u^{(n)}$ can be obtained from $e^{(1)}, \ldots, e^{(n)}$, which do form such a basis, by a unimodular transformation with matrix U. Here $e^{(k)}$ is a vector with its kth component equal to 1, and all other components zero.

Our problem may now be formulated as follows: Given $u^{(1)}$, when can we find $n - 1$ other vectors which together with $u^{(1)}$ form a basis for the lattice of all g-points?

Theorem 31 gives the answer. Consider all g-points on the line determined by $u^{(1)}$ and the origin. If, and only if, all such points are integral multiples of $u^{(1)}$, is it possible to complete $u^{(1)}$ to a basis (for the lattice of all g-points).

Let $y^{(1)}$ be a basis for all the g-points on the above straight line. Then we have $u^{(1)} = \lambda \cdot y^{(1)}$, where λ is an integer. Now, if the components of $u^{(1)}$ are relatively prime, then λ must be equal to ± 1, and $u^{(1)}$ satisfies the conditions of Theorem 31. This proves

Theorem 32. *Given n integers forming the first column of a matrix, it can be completed to a unimodular matrix if and only if the n integers are relatively prime.*

We may express this in another way: any n relatively prime integers form a *primitive vector* for the lattice of all g-points.

§3. A slight extension of Minkowski's theorem on linear forms

Theorem 29 states that there exist integers $x_1, \ldots, x_n$, not all zero, such that

$$(4) \qquad\qquad |y_j| \leq D^{1/n} \ , \quad j = 1, \ldots, n \ ,$$

where

$$(5) \qquad\qquad y_j = a_{j1} x_1 + \ldots + a_{jn} x_n \ , \quad j = 1, \ldots, n \ ,$$

and D is the absolute value of the determinant of the matrix (a_{jk}), $j, k = 1, \ldots, n$, with $D \neq 0$.

It is clear that from (4) we get a non-trivial solution of the strict inequalities

$$(6) \qquad\qquad |y_j| < c D^{1/n} \ , \quad j = 1, \ldots, n \ , \quad \text{for } c > 1 \ .$$

Let $t_1, \ldots, t_n$ be any positive numbers, such that $t_1 \ldots t_n = D$. Then if we divide y_j in (5) by t_j, for $j = 1, \ldots, n$, we get a set of n new linear forms y_j' with determinant 1 in absolute value. By Theorem 29 there exists a non-trivial solution of the inequalities

$$|y_j'| \leq 1 \ , \quad j = 1, \ldots, n \ ,$$

or

$$|y_j| \leq t_j \ , \quad j = 1, \ldots, n \ .$$

Just as before this implies the existence of a non-trivial solution of the strict inequalities

$$(7) \qquad\qquad |y_j| < c t_j \ , \quad j = 1, \ldots, n \ , \quad \text{for } c > 1 \ .$$

Let $c t_j = t_j'$. Then (7) implies the following generalization of Minkowski's theorem on linear forms.

Theorem 33. *Given n linear forms*

$$y_j = a_{j1}x_1 + \ldots + a_{jn}x_n , \quad j = 1, \ldots, n ,$$

let D be the absolute value of the determinant of the matrix (a_{jk}), $j, k = 1, \ldots, n$, with $D \neq 0$. Let $t'_1, \ldots, t'_n$ be any positive numbers, such that $t'_1 \ldots t'_n > D$. Then there exist integers $x_1, \ldots, x_n$, not all zero, such that

$$|y_j| < t'_j , \quad j = 1, \ldots, n .$$

§4. A limiting case

Note that in the last theorem, we had $t'_1 \ldots t'_n > D$. What can we say if we have $t'_1 \ldots t'_n = D$? Consider the special case of Theorem 33 for which $t'_1 = t_1(1 + \epsilon)$, $\epsilon > 0$, while $t'_k = t_k$, for $k = 2, \ldots, n$, where

$$t_1 \ldots t_n = D \neq 0 .$$

The theorem states that there exists a non-trivial solution of the inequalities

$$(8) \qquad |y_1| < t_1(1 + \epsilon) , \quad |y_2| < t_2, \quad \ldots, \quad |y_n| < t_n .$$

For a bounded ϵ, a solution of (8) lies in a bounded domain. If, for example, $0 < \epsilon < 1$, then there are only finitely many g-points $(x_1, \ldots, x_n)$ which satisfy condition (8). We can choose $\epsilon_j > 0$, such that $\epsilon_j \to 0$, as $j \to \infty$, and such that the *same* g-point, different from the origin, satisfies the inequalities in (8) for *all* ϵ_j. That g-point gives a solution of the inequalities

$$(9) \qquad |y_1| \leq t_1 , \quad |y_2| < t_2, \quad \ldots, \quad |y_n| < t_n .$$

It is clear that the sign of equality could be shifted from the first form to any of the others.

§5. A theorem about parquets

Theorems 33 and 29 depend essentially on Minkowski's First Theorem, which states that if a convex body $\mathcal{B} \subset \mathbb{R}^n$, with centre at the origin, has volume $V > 2^n$, then there exists a non-zero g-point inside $\mathcal{B}$. If $V = 2^n$, then there exists a g-point either inside $\mathcal{B}$ or on the surface $\partial\mathcal{B}$ of $\mathcal{B}$. We want to determine when the g-point is definitely inside $\mathcal{B}$.

Minkowski's First Theorem was derived from a lemma which states that if a bounded open set $\mathcal{M}$ has volume $V > 1$, then there exist two points x, y in $\mathcal{M}$ such that $x - y$ is a non-zero g-point.

The problem is now to find under what conditions this lemma holds if we merely assume that $V = 1$. In the proof of the lemma we considered sets $\mathcal{M}_g$ of the form

$$\mathcal{M}_g = \mathcal{M} + g = \{x + g | x \in \mathcal{M}\} ,$$

where g is an arbitrary integral vector. We also considered the intersections of the $\mathcal{M}_g$'s with the unit cube

$$\mathcal{E} = \{x = (x_1, \ldots, x_n) | 0 \leq x_j < 1, \quad \text{for } j = 1, \ldots, n\}.$$

There are two possibilities: *either* there is no overlapping of the intersections of $\mathcal{M}_g$ and $\mathcal{E}$, *or* at least two of the intersections overlap. In the former case, since both $\mathcal{M}$ and $\mathcal{E}$ have volume equal to 1, the set of points in $\mathcal{E}$ which are not covered by any of the $\mathcal{M}_g$'s has volume zero, and therefore it contains no interior points. We say that $\mathcal{E}$ is *filled* by the $\mathcal{M}_g$'s. In the latter case, we show just as in the original proof of the lemma that there exist two points in $\mathcal{M}$, say x and y, such that $x - y$ is a non-zero g-point.

This shows that if the points required by the lemma do *not exist*, the set $\{\mathcal{M}_g\}$, where g runs over all g-points, will fill the unit cube.

But much more is true. The whole space $\mathbb{R}^n$ is filled by the set $\{\mathcal{M}_g\}$, g running over all g-points. This follows since any point in $\mathbb{R}^n$ may be brought into $\mathcal{E}$ by a translation through an integral vector.

This consideration applied to $\mathcal{M} = \frac{1}{2}\mathcal{B}$ yields

Theorem 34. *If $\mathcal{B}$ is a convex body in $\mathbb{R}^n$ with the origin as its centre, and its volume is 2^n, there exists a non-zero g-point in $\mathcal{B}$, if and only if the set of convex bodies defined by $\frac{1}{2}\mathcal{B} + g$ do not fill $\mathbb{R}^n$, when g runs over all g-points. This happens if and only if at least two of these sets have a non-empty intersection.*

Note that $\frac{1}{2}\mathcal{B}$ is a convex body with volume 1 and centre at the origin. We denote it by $\mathcal{C}$. If the set of convex bodies $\{\mathcal{C} + g\}$ *fills* $\mathbb{R}^n$, when g runs over all g-points in $\mathbb{R}^n$, then we say that the set of convex bodies $\{\mathcal{C} + g\}$ is a *parquet* (generated by $\mathcal{C}$, with respect to the lattice of g-points. More generally we will have the obvious notion of a parquet $\{\mathcal{C} + g\}$, when g runs over all the points of an arbitrary lattice of determinant 1). It can be shown that for a parquet which arises in this way, $\mathcal{C}$ must be a polyhedron. The possible kinds of parquets are known completely for $n = 2$ and 3.

§6. Parquets formed by parallelepipeds

Let (a_{jk}) be a real matrix with determinant 1. Consider the parallelepiped defined by

$$|y_j| = |a_{j1}x_1 + \ldots + a_{jn}x_n| < 1, \quad j = 1, \ldots, n.$$

This is a convex body with the origin as centre, and volume 2^n. Let $\mathcal{C}$ be defined by the inequalities

$$|y_j| < \frac{1}{2}, \quad j = 1, \ldots, n.$$

The volume of $\mathcal{C}$ is 1. We shall investigate conditions under which the set $\{\mathcal{C} + g\}$ forms a parquet, when g runs over all g-points.

Let $y_1, \ldots, y_n$ be the coordinates of a point y. From the definition of y, it is seen that when $x = (x_1, \ldots, x_n)$ runs through all g-points, the points y lie on a lattice Λ, for which the vectors

$$a^{(k)} = (a_{1k}, \ldots, a_{nk}), \quad k = 1, \ldots, n,$$

form a basis.

Let $n = 2$. Then $\mathcal{C}$, in the x-coordinates, is a parallelogram of area 1, but in the y-coordinates it is a square, with centre at the origin, and sides (parallel to the axes) of length 1. If $\{\mathcal{C} + g\}$ is a parquet in the x-coordinates when g runs through all g-points, then since the y-coordinates are obtained by a linear transformation of the x-coordinates, $\{\mathcal{C} + g\}$ is also a parquet in the y-coordinates when g runs through all points of the lattice Λ. The set of parallelograms $\{\mathcal{C} + g\}$ transforms into the set of squares congruent to $\mathcal{C}$, whose centres are the points of the lattice Λ.

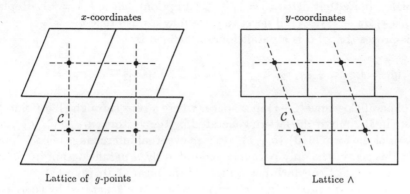

x-coordinates	y-coordinates
Lattice of g-points	Lattice Λ

Conversely a parquet in y-space will transform into a parquet in x-space. We shall show that if $\{\mathcal{C} + g\}$ is a parquet in y-space, the lattice Λ must contain either the vector $(1,0)$ or the vector $(0,1)$.

Start with the square $\mathcal{C}$, with its centre at the origin. If the set of squares $\{\mathcal{C} + g\}$, $g \in \Lambda$, forms a parquet, consider the squares which touch $\mathcal{C}$ on the side $y_1 = \frac{1}{2}$. There are two possibilities indicated by I and II. In case I, it is obvious that the vector $(1,0)$ belongs to Λ.

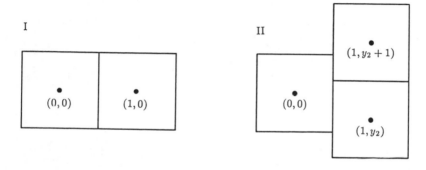

In case II, the centres of the squares abutting on the side $y_1 = \frac{1}{2}$ are the points $(1, y_2)$ and $(1, y_2 + 1)$, and the vector $(1, y_2 + 1) - (1, y_2)$ is $(0, 1)$, and this belongs to $\wedge$.

By interchanging the y_1-axis and the y_2-axis, if necessary, we may assume that $\wedge$ contains the vector $(1, 0)$. This must be a primitive vector of $\wedge$ [see § 1], for otherwise $\wedge$ would contain a vector $(\lambda, 0)$, with $0 < |\lambda| < 1$, so that $\{\mathcal{C} + g\}$, $g \in \wedge$, will not be a parquet.

Theorem 31 states that there exists a vector (b_{12}, b_{22}) which together with $(1, 0)$ form a basis for $\wedge$. This can be obtained from the basis $(a_{11}, a_{21}), (a_{12}, a_{22})$ by a unimodular transformation. Hence the absolute value of the determinant of the matrix

$$\begin{pmatrix} 1 & b_{12} \\ 0 & b_{22} \end{pmatrix}$$

also equals 1 [note that $\det(a_{jk}) = 1$, by assumption], hence $b_{22} = \pm 1$. Replacing (b_{12}, b_{22}) by $-(b_{12}, b_{22})$, if necessary, we may always take $b_{22} = +1$.

This proves that if $\mathcal{C}$ is a parallelogram defined by

$$|y_j| = |a_{j1}x_1 + a_{j2}x_2| < \frac{1}{2} , \quad j = 1, 2 , \quad a_{11}a_{22} - a_{21}a_{12} = 1 ,$$

the set of parallelograms $\{\mathcal{C} + g\}$ in x-space (where g runs through all g-points) is a parquet in $\mathbb{R}^2$ only if the lattice generated by the vectors $(a_{11}, a_{21}), (a_{12}, a_{22})$ has a basis of the form $(1, 0), (b_{12}, 1)$, after an eventual interchange of coordinate axes. In this sense, the only parquets generated by a parallelogram in $\mathbb{R}^2$ are those given by a lattice which has a basis of the form $(1, 0), (b_{12}, 1)$.

The converse was proved in Lecture VII. If a lattice generated by (a_{11}, a_{21}), (a_{12}, a_{22}) has a basis of the form $(1, 0), (b_{12}, 1)$, then there is no non-trivial solution of the inequalities $|y_1| < 1, |y_2| < 1$, in integers x_1, x_2, and Theorem 34 shows that the set of parallelograms $\{\mathcal{C} + g\}$, in x-space, forms a parquet, when g runs through all g-points.

Now let $n = 3$, and consider in y-space a cube $\mathcal{C}$ with centre at the origin, and sides (parallel to the axes) of length 1. We shall consider all cubes which touch it at the front face, and show that there must be two cubes which have a face in common, if $\{\mathcal{C} + g\}$, $g \in \wedge$, is a parquet. This implies that the vector joining their centres, which belongs to $\wedge$, is either $(1, 0, 0)$, or $(0, 1, 0)$, or $(0, 0, 1)$.

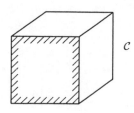

By a suitable rearrangement of the axes, we may then assume that *the lattice* Λ *always contains the vector* $(1,0,0)$.

Let F denote the face of C which lies in the plane $y_1 = \frac{1}{2}$. If there is a cube in $\{C + g\}$, $g \in \Lambda$, $g \neq 0$, with a face in common with F, the statement is proved. Otherwise we may have the following possibilities, where only the picture on the F face is given.

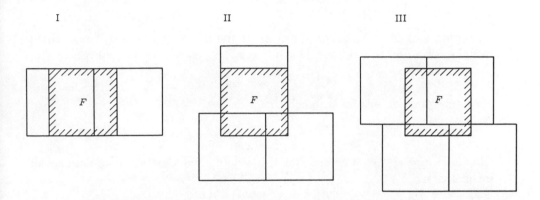

In all three cases the squares which have a side in common will be the bases of the cubes which have a face in common. This proves our statement that there will be a vector $(1,0,0)$ in the lattice Λ.

Consider an n-dimensional parquet. Suppose that we can always prove that one of the unit coordinate vectors belongs to the lattice Λ. We may take it as $(1,0,\ldots,0)$. This is a primitive vector of the lattice, for otherwise there would exist a lattice vector $(\lambda,0,\ldots,0)$, with $0 < |\lambda| < 1$. The sets C and $C + (\lambda,0,\ldots,0)$ would then overlap.

Theorem 31 shows that a basis of vectors can be found for Λ, which are column vectors of a matrix of the following form

$$
\begin{pmatrix}
1 & b_{12} & b_{13} & \cdots & b_{1n} \\
0 & b_{22} & b_{23} & \cdots & b_{2n} \\
0 & b_{32} & b_{33} & \cdots & b_{3n} \\
\vdots & \vdots & \vdots & & \vdots \\
0 & b_{n2} & b_{n3} & \cdots & b_{nn}
\end{pmatrix}.
$$

As in the case of $n = 2$, it turns out that the determinant of this matrix is ± 1. Therefore the cofactor of the element $b_{11} = 1$ equals ± 1. We can take it always as $+1$.

If $\tilde{u}_j = \sum_{k=1}^{n} b_{jk} x_k$, $j = 1,\ldots,n$, then $\tilde{y} = (\tilde{y}_1,\ldots,\tilde{y}_n)$ runs through the lattice Λ, when $x = (x_1,\ldots,x_n)$ runs through all g-points. This implies that the inequalities $|\tilde{y}_j| < 1$ have no non-trivial integral solution if and only if the same is true for the inequalities $|y_j| < 1$, $j = 1,\ldots,n$. This, by Theorem 34, happens if and only if C generates a parquet. This shows that we may take the matrix (a_{jk}) to be of the form

$$(a_{jk}) = \begin{pmatrix} 1 & a_{12} & a_{13} & \cdots & a_{1n} \\ 0 & a_{22} & a_{23} & \cdots & a_{2n} \\ 0 & a_{32} & a_{33} & \cdots & a_{3n} \\ \vdots & \vdots & \vdots & & \vdots \\ 0 & a_{n2} & a_{n3} & \cdots & a_{nn} \end{pmatrix} .$$

Consider now the linear forms

$$y_j = a_{j2}x_2 + a_{j3}x_3 + \ldots + a_{jn}x_n , \quad j = 2, \ldots, n .$$

These forms must determine a parquet in the plane $y_1 = 0$, for if not, there would exist a non-trivial integral solution $x_2, \ldots, x_n$ of the inequalities

$$|y_j| < 1 , \quad j = 2, \ldots, n .$$

Now determine an integer x_1, such that

$$-\frac{1}{2} < x_1 + a_{12}x_2 + \ldots + a_{1n}x_n \leq \frac{1}{2} .$$

This is clearly always possible, but this would show that the complete set of inequalities

$$|y_j| < 1 , \quad j = 1, \ldots, n ,$$

has a non-trivial integral solution.

By our assumption a parquet in $\mathbb{R}^{n-1}$ must contain a unit coordinate vector; therefore the matrix (a_{jk}) may be written as follows:

$$(a_{jk}) = \begin{pmatrix} 1 & a_{12} & a_{13} & \cdots & a_{1n} \\ 0 & 1 & a_{23} & \cdots & a_{2n} \\ 0 & 0 & a_{33} & \cdots & a_{3n} \\ \vdots & \vdots & \vdots & & \vdots \\ 0 & 0 & a_{n3} & \cdots & a_{nn} \end{pmatrix} .$$

By induction we may finally show the following:

The parallelepiped

$$|y_j| = |a_{j1}x_1 + a_{j2}x_2 + \ldots + a_{jn}x_n| < \frac{1}{2} , \quad j = 1, \ldots, n , \quad \det(a_{jk}) = 1 ,$$

generates a parquet in $\mathbb{R}^n$ only if the lattice defined by the vectors $(a_{11}, a_{21}, \ldots, a_{n1}), \ldots, (a_{1n}, a_{2n}, \ldots, a_{nn})$, after a suitable rearrangement of the coordinate axes, has a basis of vectors which are the columns of a triangular matrix

$$\begin{pmatrix} 1 & a_{12} & a_{13} & \cdots & a_{1n} \\ 0 & 1 & a_{23} & \cdots & a_{2n} \\ 0 & 0 & 1 & \cdots & a_{3n} \\ \vdots & \vdots & \vdots & \ddots & \vdots \\ 0 & 0 & 0 & \cdots & 1 \end{pmatrix} .$$

Note that we have proved that a parquet in $\mathbb{R}^n$ must contain a unit coordinate vector only in the cases $n = 2$ or 3. The general case is very difficult and was proved in 1941 by Hajós.

Lecture IX

§1. Products of linear forms

Consider the following n linear forms with real coefficients

$$y_j = \sum_{k=1}^{n} a_{jk} x_k \quad j = 1, \ldots, n \ ,$$

where the matrix (a_{jk}), $j, k = 1, \ldots, n$, is non-singular, and let $D = |\det(a_{jk})|$.

In Lecture VIII we proved that given n positive numbers $t_1, \ldots, t_n$ such that

$$t_1 \ldots t_n = D \ ,$$

there exists a non-zero g-point $(x_1, \ldots, x_n)$, such that

$$(1) \qquad\qquad |y_j| \leq t_j \ , \quad j = 1, \ldots, n \ .$$

This implies that there exists at least one non-trivial solution of the inequality

$$(2) \qquad\qquad |y_1 \ldots y_n| \leq D \ .$$

However, we proved more than this in Lecture III. If

$$f(y) = \frac{1}{n} \sum_{j=1}^{n} |y_j| \ , \quad y = (y_1, \ldots, y_n) \ ,$$

then f is an even gauge function on $\mathbb{R}^n$, and the volume V of the convex body $\{y | f(y) < 1\}$ is given by

$$(3) \qquad\qquad V = \frac{2^n \cdot n^n}{n!} \ .$$

Now we apply Minkowski's First Theorem. Suppose $\mu = \min f(y)$, where the minimum is taken over all points of the lattice Λ defined by (a_{jk}), other than the origin (that is, over all points $y = (y_1, \ldots, y_n)$ obtained when $(x_1, \ldots, x_n)$ runs over all g-points other than the origin). Then we know from Theorem 13 that

$$\frac{V \mu^n}{D} \leq 2^n \ ,$$

so that, on using (3), we have

(4)
$$\mu^n \le \frac{n! D}{n^n} .$$

From the inequality of the arithmetic and geometric means, we obtain

$$|y_1 \ldots y_n| \le \left(\frac{|y_1| + \ldots + |y_n|}{n} \right)^n .$$

Since by (4) the right-hand side can be made less than $\frac{n! D}{n^n}$ for some non-zero g-point, we conclude that for *this* g-point

(5)
$$|y_1 \ldots y_n| \le \frac{n! D}{n^n} .$$

If $n = 1$, the factor multiplying D is 1. If $n > 1$, the factor is less than 1, and so (5) is a better result than (2).

We shall derive the best possible result for $n = 2$, and also give the solution for $n = 3$. The best possible result for the general case is not known.

Geometrical considerations may show the difficulty of the problem. Consider the region defined by

$$\{y \,|\, |y_1 \ldots y_n| < \kappa \} , \quad \kappa > 0 .$$

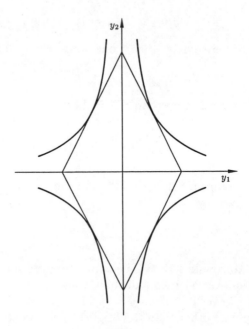

It is bounded by an n-dimensional generalization of two hyperbolas, and is obviously not convex if $n > 1$. We wish to find the minimum of the function

$$g(y) = |y_1 \cdots y_n| \,, \quad y = (y_1, \ldots, y_n) \,,$$

on the lattice Λ, without the origin. Minkowski's theorems do not apply, but if we consider the diamond-shaped regions bounded by the tangents to the hyperbolas, we have a convex region to which Minkowski's theorems do apply. Using this idea, we obtain (5). It is clear that this method will give only a very rough bound, but it is not known how to improve it in the general case.

§ 2. Product of two linear forms

We shall derive the best possible result for the case $n = 2$. We introduce a new notation. Let

$$\xi = \alpha x + \beta y \,,$$
$$\eta = \gamma x + \delta y \,,$$

and let

$$|\alpha\delta - \beta\gamma| = D > 0 \,.$$

We are going to prove that there exist integers x and y, not both zero, such that

$$|\xi\eta| \leq \frac{D}{\sqrt{5}} \,.$$

Let

$$\xi\eta = ax^2 + bxy + cy^2 = f(x,y) \,,$$

where

$$a = \alpha\gamma \,, \quad b = \alpha\delta + \beta\gamma \,, \quad c = \beta\delta \,,$$

and

$$b^2 - 4ac = (\alpha\delta - \beta\gamma)^2 = D^2 = \Delta > 0 \,.$$

We call Δ the *discriminant* of the quadratic form f. It is an indefinite quadratic form, since it can be expressed as the product of two linear forms which are linearly independent with real coefficients. [Recall that an indefinite quadratic form is a quadratic form which takes on positive as well as negative values.] Conversely, an indefinite binary quadratic form may be expressed as the product of two linear forms which are linearly independent with real coefficients, and therefore its discriminant is positive.

We now state

Theorem 35. *Let f be an indefinite binary quadratic form with discriminant Δ. Then there exists a g-point (x,y), other than the origin, such that*

$$|f(x,y)| \leq \sqrt{\frac{\Delta}{5}} \,.$$

In other words, the greatest lower bound of $|f|$ on the lattice of g-points, excluding the origin, is not greater than $\sqrt{\frac{\Delta}{5}}$.

To prove Theorem 35 we use the fact that a unimodular transformation of $\mathbb{R}^2$ leaves the origin and the set of g-points unchanged. Consider the unimodular transformation U defined by

$$U : \begin{cases} x = px_1 + qy_1 \ , \\ y = rx_1 + sy_1 \ , \end{cases}$$

where p, q, r, s are integers, such that $ps - qr = \pm 1$. Suppose U transforms $f(x, y)$ into $f_1(x_1, y_1)$, where

$$f_1(x_1, y_1) = a_1 x_1^2 + b_1 x_1 y_1 + c_1 y_1^2 \ ,$$

then $b_1^2 - 4a_1 c_1 = \Delta$. The set of values taken by $f_1(x_1, y_1)$ as (x_1, y_1) runs over all g-points, will be the same as the set of values taken by $f(x, y)$ as (x, y) runs over all g-points, since both U and U^{-1} are unimodular transformations.

Two forms, such as f and f_1, which are related by a unimodular transformation are called *equivalent*. We use the notation $f \sim f_1$ to express the fact that f is equivalent to f_1. This concept will help us to reduce our problem to the consideration of a simpler one. For example, the discussion of the minimum of

$$f(x, y) = 7x^2 + 8xy + 2y^2$$

is reduced by the transformation

$$U : \begin{cases} x = -x_1 - 2y_1 \ , \\ y = x_1 + 3y_1 \ , \end{cases}$$

to the discussion of the minimum of

$$f_1(x_1, y_1) = x_1^2 - 2y_1^2 \ .$$

It is obvious that the minimum of $|f_1|$ on the lattice of g-points excluding the origin is 1. Therefore the minimum value of $|f|$ is also 1.

We may sharpen the statement of Theorem 35 as follows:

Theorem 35′. *Let f be an indefinite binary quadratic form with discriminant Δ. Then either there exists a g-point (x, y), different from the origin, such that*

$$|f(x, y)| < \sqrt{\frac{\Delta}{5}} \ ,$$

or f is equivalent to the form $m(x^2 - xy - y^2)$, with $m \neq 0$. In the latter case, there exists a solution of the equation

$$|f(x, y)| = \sqrt{\frac{\Delta}{5}} = |m| \ ,$$

in integers x, y, and $|m|$ is the minimum of $|f(x, y)|$ over all g-points (x, y) different from the origin.

It is clear that $|m|$ *is* the minimum of $|f(x,y)| = |m(x^2 - xy - y^2)|$, since there does not exist a non-zero g-point (x,y), for which $|x^2 - xy - y^2| < 1$.

Note that the 'extremal form' $x^2 - xy - y^2$ is equivalent to the form $x^2 + xy - y^2$, for instance by the unimodular transformation $(x,y) \mapsto (x+y,y)$. The further unimodular transformation $(x,y) \mapsto (y,x)$ shows that the forms $x^2 + xy - y^2$ and $-(x^2 - xy - y^2)$ are equivalent. Hence the form $m(x^2 - xy - y^2)$ is equivalent to the form $-m(x^2 - xy - y^2)$; it follows that in the second alternative in Theorem 35′, we can assume that $m > 0$.

We may sharpen the theorem still further:

Theorem 35″. *If the indefinite binary quadratic form $f = \xi\eta$, with discriminant Δ, is not equivalent to $m(x^2 - xy - y^2)$, with $m > 0$, then there exists a non-trivial g-point (x,y), such that*

$$|\xi\eta| = |f(x,y)| < \sqrt{\frac{\Delta}{5}} \ ,$$

and such that for an arbitrarily chosen $\epsilon > 0$, we have

$$|\xi| < \epsilon.$$

If f is equivalent to $m(x^2 - xy - y^2)$, there exists an integral solution of the equation

$$f(x,y) = m \ ,$$

such that we have, in addition, $|\xi| < \epsilon$, for any preassigned $\epsilon > 0$.

This theorem implies that there exists an infinite number of pairs of integers (x,y), such that

$$|f(x,y)| \leq \sqrt{\frac{\Delta}{5}} \ .$$

For if there exists a non-zero integral solution (x,y) for which we have $\xi = 0$, then $f(x,y) = 0$, and hence also $f(jx,jy) = 0$, for all integral j, since f is homogeneous in x and y, giving thus an infinite number of solutions. If, on the other hand, the solution (given by the theorem) is such that the corresponding value of ξ, say ξ^*, is different from zero, we can take ϵ_1, with $0 < \epsilon_1 < |\xi^*|$, and obtain (by the theorem) a new solution, say (x_1, y_1). Again for the new solution we have either $\xi = 0$, in which case there is an infinity of solutions, or we can argue with another ϵ_2, and so on.

Proof of Theorem 35″. Consider the values of $|f(x,y)|$ on the lattice of g-points excluding the origin. This set has a greatest lower bound, say m. We may assume that $m > 0$. Either m is attained at some g-point, or values arbitrarily close to m are attained. In any case, there exist integers p, r and a real number k, $\frac{1}{2} < k \leq 1$, such that

(6) $$|f(p,r)| = \frac{m}{k} \ .$$

Let h denote the greatest common divisor of p and r. If $h > 1$, we have

$$\left| f\left(\frac{p}{h}, \frac{r}{h}\right) \right| = \frac{m}{kh^2} < \frac{m}{2} ,$$

since h is an integer and must therefore be equal to or greater than 2. This contradicts our assumption that m is the greatest lower bound of $|f(x,y)|$. Therefore $h = 1$.

We proved in Lecture VIII that any column vector of relatively prime integers may be completed to form a unimodular matrix. Suppose this is done for p, r. Let $\begin{pmatrix} p & q \\ r & s \end{pmatrix}$ be a unimodular matrix. Consider the unimodular transformation $(x_1, y_1) \mapsto (x, y)$ given by

$$x = px_1 + qy_1 ,$$
$$y = rx_1 + sy_1 .$$

Suppose that $f(x,y)$ becomes $f_1(x_1, y_1) = a_1 x_1^2 + b_1 x_1 y_1 + c_1 y_1^2$ under this transformation; ξ becomes $\xi_1 = \alpha_1 x_1 + \beta_1 y_1$, and η becomes $\eta_1 = \gamma_1 x_1 + \delta_1 y_1$. Note that the point $(x_1, y_1) = (1, 0)$ goes over into the point $(x, y) = (p, r)$. We have therefore

$$\frac{m}{k} = |f_1(1, 0)| = |a_1| = |\alpha_1 \gamma_1| .$$

Note that $\alpha_1 \gamma_1 \neq 0$, since $m > 0$ by assumption. If $\alpha_1 \gamma_1 < 0$, we may replace η by $-\eta$, and f by $-f$, and then apply the unimodular transformation defined above, so that $\alpha_1 \gamma_1$ will be positive. [Recall that we have shown that $m(x^2 - xy - y^2) \sim -m(x^2 - xy - y^2)$.] Therefore we may restrict ourselves to forms f_1 with $\alpha_1 \gamma_1 > 0$.

We may also assume that $\alpha_1 > 0$. If not, since $\alpha_1 \gamma_1 > 0$, we must also have $\gamma_1 < 0$. Then replace (x_1, y_1) by $(-x_1, y_1)$, and we shall have $\alpha_1 > 0$, $\gamma_1 > 0$.

We may take β_1 so that $-\alpha_1 \leq \beta_1 < 0$. For consider the unimodular transformation

$$x_1 = x_2 + ly_2 ,$$
$$y_1 = \qquad y_2 ,$$

where l is an integer, under which $\xi_1 = \alpha_1 x_1 + \beta_1 y_1$ goes over into $\xi_2 = \alpha_2 x_2 + \beta_2 y_2 = \alpha_1 x_2 + (l\alpha_1 + \beta_1)y_2$, and η_1 goes over into $\eta_2 = \gamma_2 x_2 + \delta_2 y_2$, where $\gamma_2 = \gamma_1$. Choose l to be the smallest integer which is greater than or equal to $-1 - \frac{\beta_1}{\alpha_1}$. Then we have

$$-\alpha_1 \leq l\alpha_1 + \beta_1 < 0 ,$$

or

$$-\alpha_2 \leq \beta_2 < 0 .$$

We may also suppose that $\alpha_2 \delta_2 - \beta_2 \gamma_2 = +D$ [recall $(\alpha_2 \delta_2 - \beta_2 \gamma_2)^2 = D^2$], since otherwise we can replace y_2 by $-y_2$.

To sum up: given any indefinite binary quadratic form

$$f(x, y) = \xi\eta = (\alpha x + \beta y)(\gamma x + \delta y) , \quad \text{with } \alpha\gamma > 0 ,$$

by suitable unimodular transformations we can always make it satisfy the following conditions:

(i) $|f(1,0)| = \alpha\gamma = \dfrac{m}{k}$,

(ii) $\alpha > 0$, $\gamma > 0$, $-\alpha \leq \beta < 0$;

(iii) $\alpha\delta - \beta\gamma = D > 0$.

Since $m > 0$ by assumption, the case $-\alpha = \beta$ in condition (ii) may be excluded, for if $-\alpha = \beta$, then $\xi = 0$ for $x = 1$, $y = 1$, which implies that $m = 0$, in contradiction to the assumption $m > 0$.

Geometrically these conditions amount to taking one asymptote of the hyperbola $f(x,y) =$ a constant, with a slope greater than 1, and making sure that the function $(x,y) \mapsto |f(x,y)|$ takes a value fairly close to m at the point $(1,0)$.

Exactly the same considerations apply, if we assume that m is now the greatest lower bound of $|f(x,y)|$, when (x,y) runs over all g-points, other than the origin, which satisfy the further condition that

(7) $$|\xi| = |\alpha x + \beta y| < \epsilon .$$

[Obviously the theorem is trivial in the case $m = 0$, which justifies the assumption that $m > 0$.] Since this has to be satisfied by the point $(1,0)$, condition (ii) implies condition

(iv) $0 < \alpha < \epsilon$.

Thus, from the definition of m, we have

$$|(\alpha x + \beta y)(\gamma x + \delta y)| \geq m ,$$

for all integers x, y, not both equal to zero, which satisfy condition (7). Dividing by $\alpha\gamma$, which is positive, we have

$$\left|\left(x + \frac{\beta}{\alpha}y\right)\left(x + \frac{\delta}{\gamma}y\right)\right| \geq \frac{m}{\alpha\gamma} = k , \quad \text{by (i)} ,$$

or if we set

$$\frac{\beta}{\alpha} = -\lambda , \quad \frac{\delta}{\gamma} = -\mu ,$$

we obtain

(8) $$|(x - \lambda y)(x - \mu y)| \geq k .$$

Note that conditions (i) and (iii) imply that

(9) $$\lambda - \mu = \frac{Dk}{m} ,$$

while condition (7) becomes

(10) $|x - \lambda y| < \dfrac{\epsilon}{\alpha}$,

and condition (ii), with $-\alpha \neq \beta$, implies that

(11) $0 < \lambda < 1$.

We shall now show that from (8), (9), (10), and (11), it follows that

$$m \leq \frac{D}{\sqrt 5} .$$

Since $\epsilon > \alpha$, by (iv), the points (0,1) and (1,1) satisfy the condition in (10) because of (11). Therefore from (8) we get

(12) $|\lambda\mu| \geq k$,

and

(13) $|(1 - \lambda)(1 - \mu)| \geq k$.

Suppose now that $\mu \geq 0$. From (12) it follows that $\mu > 0$, and from (9) that $\mu < \lambda$, hence from (11) that $0 < \mu < 1$. But then $\lambda\mu > 0$, and $(1-\lambda)(1-\mu) > 0$, and multiplying inequalities (12) and (13), we get

$$\lambda(1 - \lambda)\mu(1 - \mu) \geq k^2 > \frac{1}{4} ,$$

since $\frac{1}{2} < k$. This is impossible, because for any real v, we have $(v - \frac{1}{2})^2 = v^2 - v + \frac{1}{4} \geq 0$, or $v(1 - v) \leq \frac{1}{4}$, and therefore

$$\lambda(1 - \lambda)\mu(1 - \mu) \leq \frac{1}{16} .$$

Hence $\mu < 0$, and we may write (12) and (13) as

$$-\lambda\mu \geq k ,$$

and

$$(1 - \lambda)(1 - \mu) \geq k ,$$

from which it follows that

(14) $1 - (\lambda + \mu) - k \geq -\lambda\mu \geq k$,

or, since $k > \frac{1}{2}$,

$$-(\lambda + \mu) \geq 2k - 1 > 0 .$$

But we have, from (9),

(15) $\dfrac{D^2 k^2}{m^2} = (\lambda - \mu)^2 = (\lambda + \mu)^2 - 4\lambda\mu \geq (2k - 1)^2 + 4k = 4k^2 + 1$,

and therefore

(16)
$$m^2 \le \frac{D^2}{4 + \left(\frac{1}{k^2}\right)} \le \frac{D^2}{5} = \frac{\Delta}{5} \, ,$$

since k is at most 1. Hence we have *either* $m < \sqrt{\frac{\Delta}{5}}$, *or* $m = \sqrt{\frac{\Delta}{5}}$.

It remains to be shown that, in the latter case, f is equivalent to $m(x^2 - xy - y^2)$. The equality $m = \sqrt{\frac{\Delta}{5}}$ is possible only if $k = 1$, and if in (14) equality takes the place of the inequality, in which case we have $-(\lambda + \mu) = -\lambda\mu = 1$, or $\alpha\gamma = \alpha\delta + \beta\gamma = -\beta\delta$, and since $\alpha\gamma = \frac{m}{k} = m$, we have $f(x, y) = m(x^2 + xy - y^2)$, which is equivalent to $m(x^2 - xy - y^2)$.

§3. Approximation of irrationals

Let ω be a positive irrational number. Consider the quadratic form

$$(x - \omega y)y = xy - \omega y^2 \, .$$

Its discriminant is 1; therefore, by Theorem 35'', there exist two integers x and y, not both zero, such that

(17)
$$|(x - \omega y)y| \le \frac{1}{\sqrt{5}} \, .$$

The equality can occur only if $xy - \omega y^2$ is equivalent to $m(x^2 - xy - y^2)$. But since ω is irrational, this equivalence is impossible, and the strict inequality must hold. Furthermore we can obtain non-trivial solutions of (17) under the additional condition

(18)
$$|x - \omega y| < \epsilon \, , \quad \epsilon > 0 \, .$$

Note that, for $0 < \epsilon \le 1$, y cannot be zero, for then (18) would imply that x is also zero. Dividing by y^2, we find

Theorem 36. *Let ω be a positive irrational number. Then there exist infinitely many pairs of integers x, y, such that*

$$\left| \frac{x}{y} - \omega \right| < \frac{1}{\sqrt{5} \cdot y^2} \, .$$

The infinite number of solutions is a consequence of (18), since ϵ can be chosen arbitrarily small.

The constant $\frac{1}{\sqrt{5}}$ in the theorem is the best possible, since it cannot be improved for $\omega = \frac{1 + \sqrt{5}}{2}$.

The result of this theorem was first obtained by Hurwitz with the help of continued fractions.

§4. Product of three linear forms

Let

$$y_j = \sum_{k=1}^{3} a_{jk} x_k , \quad j = 1, 2, 3 ,$$

and $D = |\det(a_{jk})| \neq 0$. Then from (5), with $n = 3$, we deduce that there exists a non-trivial solution in integers x_1, x_2, x_3 of the inequality

$$|y_1 y_2 y_3| \leq \frac{2}{9} D .$$

This result can be improved. Davenport found that there always exists a non-trivial solution of the inequality

$$|y_1 y_2 y_3| < \frac{D}{7} ,$$

unless the product $y_1 y_2 y_3$ is equivalent to

$$\frac{D}{7} \left(x_1 + 2\cos\frac{2\pi}{7} x_2 + 2\cos\frac{4\pi}{7} x_3 \right) \left(x_1 + 2\cos\frac{4\pi}{7} x_2 + 2\cos\frac{6\pi}{7} x_3 \right)$$

$$\cdot \left(x_1 + 2\cos\frac{6\pi}{7} x_2 + 2\cos\frac{2\pi}{7} x_3 \right) ,$$

in which case there exists a solution of the equality

$$y_1 y_2 y_3 = \frac{D}{7} .$$

Note that the extremal form in Theorem 35'' may be written as

$$y_1 y_2 = \frac{D}{\sqrt{5}} \left(x_1 + 2\cos\frac{2\pi}{5} x_2 \right) \left(x_1 + 2\cos\frac{4\pi}{5} x_2 \right) .$$

From these two cases we might conjecture that there always exists a non-trivial solution in integers of the inequality

$$|y_1 \ldots y_n| \leq KD ,$$

where $y_j = \sum_{k=1}^{n} a_{jk} x_k$, $j = 1, \ldots, n$, and $D = |\det(a_{jk})|$, $K = (2n+1)^{(1-n)/2}$. The statement is true for $n = 1, 2, 3$, but for $n = 4$ the extremal form would seem to be the product of factors such as

$$x_1 + 2\cos\frac{2\pi}{9} x_2 + 2\cos\frac{4\pi}{9} x_3 + 2\cos\frac{6\pi}{9} x_4 .$$

But for $x_1 = x_4 = 1$, $x_2 = x_3 = 0$, this factor becomes zero. It is clear that the statement is not true in general. However if $2n + 1 = p$, a prime, it can be proved that K cannot be smaller than $(2n + 1)^{(1-n)/2}$.

§5. Minimum of positive-definite quadratic forms

Theorem 37. *Let $f(x,y) = ax^2 + bxy + cy^2$ be a positive-definite quadratic form, and let $4ac - b^2 = \Delta$ [note the change in sign, when compared with the notation in § 2]. Then $f(x,y)$ attains a minimum on the lattice of g-points (x,y) other than the origin, and the minimum is not greater than $\sqrt{\frac{\Delta}{3}}$. The minimum equals $\sqrt{\frac{\Delta}{3}}$, if and only if $f(x,y)$ is equivalent to the form $m(x^2 + xy + y^2)$, for an $m > 0$.*

Since f is positive-definite, there exists a linear transformation given by

$$\begin{cases} x = \alpha x_1 + \beta y_1 , \\ y = \gamma x_1 + \delta y_1 , \end{cases}$$

which carries $f(x,y)$ into $f_1(x_1, y_1) = x_1^2 + y_1^2$. The set $\{(x_1, y_1) | x_1^2 + y_1^2 < 1\}$ is a convex body with the origin as centre; so is the set $\{(x,y) | [f(x,y)]^{1/2} < 1\}$. If $\lambda > 0$, we have $[f(\lambda x, \lambda y)]^{1/2} = \lambda [f(x,y)]^{1/2}$. Hence $f^{1/2}$ is a gauge function. Let m denote the minimum of $f(x,y)$ over all g-points (x,y) different from the origin. Then there exists a g-point (p,r), different from the origin, such that $[f(p,r)]^{1/2} = m^{1/2}$. As in the case of indefinite forms, the column vector $\begin{pmatrix} p \\ r \end{pmatrix}$ can be completed to a unimodular matrix. And the corresponding unimodular transformation carries the point $(1,0)$ into the point (p,r), and carries f into a new form which attains the value m at the point $(1,0)$. We may therefore assume that for the transformed positive-definite form (denoted again by)

$$f(x,y) = ax^2 + bxy + cy^2 ,$$

we have $f(1,0) = a = m$. Note that $m \neq 0$, since f is positive-definite. Since m is the minimum of f, and $f(0,1) = c$, we must have $a \leq c$. Just as before, we may suppose that $0 \leq b \leq a$. For the unimodular transformation

$$\begin{cases} x = x_1 \pm ly_1 , \\ y = \pm y_1 , \end{cases}$$

transforms f into a form f_1, say, in which the coefficient of $x_1 y_1$ is $\pm(b + 2al)$. We can always find an integer l, such that

$$-a \leq b + 2al < a .$$

We may therefore finally assume that

$$0 \leq b \leq a \leq c , \quad a = m .$$

Now

$$\frac{\Delta}{m^2} = \frac{4ac - b^2}{a^2} = \frac{4c}{a} - \left(\frac{b}{a}\right)^2 \geq 4 - 1 = 3 ;$$

therefore

(19)
$$m \le \sqrt{\frac{\Delta}{3}} \; .$$

For equality here we must have

$$\frac{c}{a} = 1 \;, \qquad \frac{b}{a} = 1 \;,$$

that is, $f(x,y)$ must be equivalent to $m(x^2 + xy + y^2)$. If f is equivalent to such a form, then $\Delta = 3m^2$. This completes the proof of Theorem 37.

Note that this theorem improves Minkowski's First Theorem for elliptic discs in $\mathbb{R}^2$. Consider the convex body bounded by the ellipse given by the equation

$$f(x,y) = 1 \;, \quad \text{where } f(x,y) = ax^2 + bxy + cy^2 \;, \quad \text{with}$$

$$\Delta = 4ac - b^2 > 0 \;.$$

The corresponding gauge function is $f^{1/2}$, and the convex body $\{(x,y)|[f(x,y)]^{1/2} < 1\}$ is an elliptic disc with volume $V = \frac{2\pi}{\sqrt{\Delta}}$. If m denotes the minimum of $f(x,y)$, and μ that of $[f(x,y)]^{1/2}$, over all g-points different from the origin, we get from Theorem 12, which is an immediate consequence of Minkowski's First Theorem,

$$V \mu^2 \le 4 \;.$$

Since $m = \mu^2$, $V = \frac{2\pi}{\sqrt{\Delta}}$, we can rewrite this as $m \le \frac{2\sqrt{\Delta}}{\pi}$. But we have proved that $\frac{1}{\sqrt{3}}$ is the best constant on the right-hand side of (19), therefore $\frac{1}{\sqrt{3}} \le \frac{2}{\pi}$!

Chapter III

Theory of Reduction

Lectures X to XV

Lecture X

§1. The problem of reduction

In the two previous lectures we have obtained theorems about the minima of quadratic forms on the set of all non-zero integral points. These theorems can be formulated in terms of lattices. For example, the theorem about the minimum of positive-definite binary quadratic forms gives the following result:

Any lattice in $\mathbb{R}^2$ of rank 2 with determinant D contains a non-zero point (X, Y), such that

$$X^2 + Y^2 \leq \frac{2D}{\sqrt{3}} .$$

If we wished to check these results, we would have to consider all lattices defined by the formulae

(1) $$X = \alpha x + \beta y , \quad Y = \gamma x + \delta y ,$$

where x, y are integers, and $|\alpha\delta - \beta\gamma| = D$.

It is clear, however, that definition (1) will include each lattice an infinite number of times, for any unimodular transformation will leave the lattice unchanged as a geometrical entity, but will change its analytic definition. The diagrams illustrate this. The lattices defined by

$$X = x, Y = y , \quad \text{and } X = x, Y = 2x + y$$

are seen to be identical. [See next page.]

It would be advantageous to pick out one lattice from the class of all lattices which are equivalent under a unimodular transformation. This lattice would then represent the entire class of equivalent lattices.

The problem of finding such a representative for every class of equivalent lattices is called the problem of *reduction*. A major part of the theory of reduction is due to Minkowski, but there have been important developments in recent (?) years. The theory has important applications to the study of algebraic functions of several variables, differential equations, and geometry.

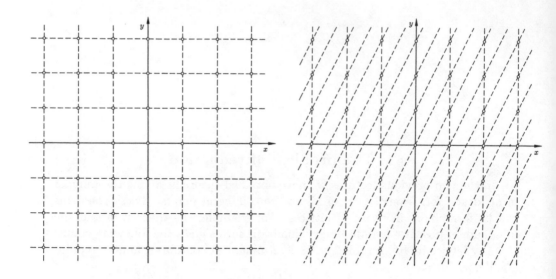

§2. Space of all matrices

Let $\wedge$ be a lattice in $\mathbb{R}^n$ of rank n. It has a *basis*, that is, n linearly independent vectors $a^{(1)}, \ldots, a^{(n)}$, such that if x is any vector in $\wedge$, then x can be expressed as $g_1 a^{(1)} + \ldots + g_n a^{(n)}$, where $g_1, \ldots, g_n$ are integers.

Consider the vectors $a^{(j)}$, $j = 1, \ldots, n$, as the column vectors of a matrix A, that is

$$A = \begin{pmatrix} a_1^{(1)} & \cdots & a_1^{(n)} \\ \vdots & & \vdots \\ a_n^{(1)} & \cdots & a_n^{(n)} \end{pmatrix}.$$

We know that the basis is not unique, and that any two bases for the lattice $\wedge$ are connected by a unimodular transformation U [Lecture V, Theorem 19]. If we let B represent the matrix of a new basis, then $B = AU$.

We may also go *from* a matrix *to* a lattice by starting with an arbitrary matrix of non-zero determinant. Let A be such a matrix, and $a^{(j)}$, $j = 1, \ldots, n$, its column vectors. Then there exists a lattice $\wedge$, such that the vectors $a^{(j)}$, $j = 1, \ldots, n$, are a basis for it. It is clear that $\wedge$ is composed of all vectors of the form $g_1 a^{(1)} + \ldots + g_n a^{(n)}$, where $g_1, \ldots, g_n$ are integers.

A geometrical interpretation of the relationship between matrices and lattices will be useful. We proceed as follows:

The n^2 components of a non-singular matrix may be considered as the n^2 coordinates of a point in n^2-dimensional Euclidean space. We shall denote this *matrix space* by $\mathfrak{A}$. [According to the notation widely in use nowadays, $\mathfrak{A} = GL(n, \mathbb{R}) \subset gl(n, \mathbb{R}) \cong \mathbb{R}^{n^2}$.] To every non-singular matrix corresponds a unique point in $\mathfrak{A}$, and conversely to every point in $\mathfrak{A}$ corresponds a unique non-singular matrix. We have just seen that to each non-singular matrix corresponds a unique lattice Λ, but that to every lattice correspond an infinite number of matrices equivalent under unimodular transformation, for $n \geq 2$. In geometrical language we may say that to each point in $\mathfrak{A}$, there corresponds a lattice, but to each lattice there correspond an infinite number of points of $\mathfrak{A}$.

The space $\mathfrak{A}$ is a *group-space*; if A and C are in $\mathfrak{A}$, then AC and A^{-1} are also in $\mathfrak{A}$. In this group there exists a discrete subgroup, the group of all unimodular matrices U.

The set of points $\{AU\}$, where $A \in \mathfrak{A}$ is some fixed matrix, and U runs through the group of unimodular matrices, is called a *left coset* of the group space with respect to the unimodular group. Then to every lattice Λ there is associated a left coset.

Our problem is to pick out one and only one element from this coset to represent the lattice. Of course this can be done in many ways, but we would prefer a *method* which has a simple geometric interpretation: for example, we might require the basis vectors to have certain *extremal* properties.

§3. Minimizing vectors

A possible method for defining this basis would seem to be given by Minkowski's Second Theorem. Let us recall the details.

Given a lattice Λ in $\mathbb{R}^n$ of rank n and an arbitrary even gauge function f on $\mathbb{R}^n$, there exist n linearly independent vectors $a^{(1)}, \ldots, a^{(n)}$ in Λ, such that

$$(2) \qquad f\left(a^{(k)}\right) = \mu_k , \quad k = 1, \ldots, n ,$$

where μ_k is the kth minimum of f on the lattice Λ. Let V be the volume of the convex body $\{x | f(x) < 1\}$. Let D be the determinant of Λ, and let Θ be the absolute value of the determinant of the $a^{(k)}$, $k = 1, \ldots, n$. Then Minkowski's Second Theorem [see Lecture III, Theorem 16; and Lecture IV, § 5] states that

$$(3) \qquad \frac{2^n \Theta}{n! D} \leq \frac{V}{D} \cdot \mu_1 \ldots \mu_n \leq 2^n .$$

If the vectors $a^{(k)}$, $k = 1, \ldots, n$, constitute a basis for Λ, we would have a basis defined by extremal properties, and if the basis is unique, our problem will be solved. In general, for $n > 2$, it is not true that the vectors $a^{(k)}$, $k = 1, \ldots, n$, form a basis for Λ, nor are they unique.

Let us consider the method by which the vectors $a^{(k)}$ were defined. Given the vectors $a^{(1)}, \ldots, a^{(k-1)}$, we consider the values of f on the set of all lattice vectors x linearly independent of $a^{(1)}, \ldots, a^{(k-1)}$. We showed before that f has a minimum on this set. The kth *minimizing vector* of f is a vector $a^{(k)}$ at which this minimum is attained, so that we have

$$f(x) \geq f\left(a^{(k)}\right) ,$$

for all lattice vectors x linearly independent of $a^{(1)}, \ldots, a^{(k-1)}$. Suppose we had arranged it so that the vectors $a^{(1)}, \ldots, a^{(k-1)}$ form a basis for the lattice of all vectors in $\wedge$ which lie in the linear manifold, through the origin in $\mathbb{R}^n$, generated by $a^{(1)}, \ldots, a^{(k-1)}$, then we could not be sure that the vectors $a^{(1)}, \ldots, a^{(k)}$ would have the same property.

§4. Primitive sets

We can see that we must modify our procedure so that, while still retaining the extremal properties, we could be sure that we always have a basis. This property is already familiar to us in the idea of primitive sets [Lecture VIII, § 1]. A set of vectors $c^{(1)}, \ldots, c^{(k)}$ in $\wedge$ is a *primitive set*, if whenever any linear combination

$$\lambda_1 c^{(1)} + \ldots + \lambda_k c^{(k)} ,$$

with real λ_j, $j = 1, \ldots, k$, is a lattice vector, then the numbers λ_j, $j = 1, \ldots, k$, are integers. We shall prove again [cf. Theorem 31] the following

Lemma 1. *Any primitive set of vectors in a lattice $\wedge$ in $\mathbb{R}^n$ of rank n can be completed to a basis for the lattice.*

Let $c^{(1)}, \ldots, c^{(k)}$ be a primitive set, and let $d^{(k+1)}, \ldots, d^{(n)}$ be $n - k$ vectors in $\wedge$, such that the set of vectors

$$c^{(1)}, \ldots, c^{(k)} , \quad d^{(k+1)}, \ldots, d^{(n)}$$

are linearly independent. We have proved [Lecture V, Theorem 18] that a basis $b^{(1)}, \ldots, b^{(n)}$ can be constructed from these n vectors, so that

$$b^{(i)} = \sum_{j=1}^{i-1} r_{ij} c^{(j)} + r_i c^{(i)} , \qquad\qquad i = 1, \ldots, k ;$$

and

$$b^{(i)} = \sum_{j=1}^{k} r_{ij} c^{(j)} + \sum_{j=k+1}^{i-1} r_{ij} d^{(j)} + r_i d^{(i)} , \quad i = k+1, \ldots, n ,$$

where the r_{ij} are rational numbers, and r_i^{-1} are positive integers.

Since the $c^{(i)}$, $i = 1, \ldots, k$, form a primitive set, it follows that $r_i = 1$, $r_{ij} =$ an integer, for $i = 1, \ldots, k$; $j = 1, \ldots, i - 1$.

This shows that $b^{(1)} = c^{(1)}$. Now if $b^{(i)}$ is replaced by $b^{(i)}$ minus any integral linear combination of the preceding basis vectors $b^{(1)}, \ldots, b^{(i-1)}$, we still have a basis for $\wedge$. In particular, we may replace $b^{(2)}$ by $b^{(2)} - r_{21} b^{(1)} = c^{(2)}$. In the same way, we may replace $b^{(3)}$ by $c^{(3)}$, and so on, until $b^{(k)}$ is replaced by $c^{(k)}$. This shows that the set

$$c^{(1)}, \ldots, c^{(k)}, b^{(k+1)}, \ldots, b^{(n)}$$

is a basis for $\wedge$, and proves the lemma.

Note that since every part of a basis is a primitive set, the set

$$c^{(1)}, \ldots, c^{(k)}, b^{(k+1)}$$

is also primitive. This shows that given a set of k primitive vectors, where $k < n$, it can always be extended to a set of $k + 1$ primitive vectors.

§5. Construction of a reduced basis

The method used before to obtain the minimizing vectors $a^{(k)}$, $k = 1, \ldots, n$, will now be applied to *primitive* sets of vectors instead of to sets of *linearly independent* vectors.

Consider all primitive sets of one vector b. Pick $b^{(1)}$, such that $f\left(b^{(1)}\right)$ is a minimum, equal to λ_1, say. Note that $a^{(1)}$ [as in § 3] is necessarily primitive, so that $\lambda_1 = \mu_1$. Consider all b such that $b^{(1)}$, b is primitive. Choose b such that $b = b^{(2)}$, say, which makes $f\left(b^{(2)}\right)$ a minimum, equal to λ_2. Continue in the same way. Let $b^{(1)}, \ldots, b^{(k-1)}$ be a primitive set. Consider all b such that $b^{(1)}, \ldots, b^{(k-1)}$, b form a primitive set. Take $b^{(k)}$ among these b, for which $f\left(b^{(k)}\right)$ is a minimum, equal to λ_k. In this way we get a primitive set of n vectors, that is, a basis for the given lattice.

A numerical illustration may clarify the method. Consider the lattice of g-points in $\mathbb{R}^3$, and the gauge function f defined by

$$f(x) = |2x_1 - x_3| + |2x_2 - x_3| + |x_3|, \quad x = (x_1, x_2, x_3).$$

It can be easily shown that $\mu_1 = \mu_2 = \mu_3 = 2$, with minimizing vectors $(1,0,0)$, $(0,1,0)$, and $(1,1,2)$. To *reduce* the lattice by means of this gauge function, we may start with $b^{(1)} = (1,0,0)$ and $\lambda_1 = 2$. For $b^{(2)}$ we may take either $(0,1,0)$, or $(1,1,2)$, or their negatives. In either case, we have $\lambda_2 = 2$. Suppose we take $b^{(2)} = (0,1,0)$. Then $b^{(3)}$ may be, for example, $(0,0,1)$ or $(1,1,1)$, and $\lambda_3 = 3$. Notice that the basis vectors are not unique, for we may have, for example, $(1,0,0)$, $(0,1,0)$, $(0,0,1)$ or $(1,0,0)$, $(0,1,0)$, $(1,1,1)$ as the basis vectors.

It is clear that λ_1 must equal μ_1, but otherwise there is no immediate relation between the values of λ_k and μ_k (other than the trivial one: $\mu_k \leq \lambda_k$). We shall show, however, that the λ_k satisfy an inequality similar to that in Minkowski's Second Theorem. We call it the First Finiteness Theorem.

§6. The First Finiteness Theorem

For any even gauge function f on $\mathbb{R}^n$, and any lattice Λ in $\mathbb{R}^n$ of rank n, we have

(4)
$$\frac{2^n}{n!} \leq \frac{V}{D} \cdot \lambda_1 \ldots \lambda_n \leq 2^n \cdot \left(\frac{3}{2}\right)^{n(n-1)/2},$$

where V is the volume of the convex body $\{x | f(x) < 1\}$, $\lambda_1, \ldots, \lambda_n$ are defined as above, and D is the determinant of the lattice Λ.

The left-hand inequality in (4) is trivial. It expresses the fact that the closure of the convex body just defined by $\{x \mid f(x) < 1\}$ contains the points

$$\pm \frac{1}{\lambda_1} b^{(1)}, \ldots, \pm \frac{1}{\lambda_n} b^{(n)} ,$$

so that V is not less than the volume of the n-dimensional octahedron whose vertices are $\pm \frac{1}{\lambda_1} b^{(1)}, \ldots, \pm \frac{1}{\lambda_n} b^{(n)}$. The linear transformation which transforms $\frac{1}{\lambda_k} b^{(k)}$ into the kth unit vector $e^{(k)}$, will transform the octahedron whose vertices are $\pm \frac{1}{\lambda_k} b^{(k)}$ into the octahedron defined by

$$|y_1| + \ldots + |y_n| \leq 1 .$$

The volume of this octahedron is $\frac{2^n}{n!}$. The volume of the original octahedron will be $\frac{2^n}{n!}$ times the absolute value of the determinant of the linear transformation which takes $e^{(k)}$ into $\frac{1}{\lambda_k} b^{(k)}$. The absolute value of the determinant of this transformation is

$$\frac{\left| \det \left(b^{(1)}, \ldots, b^{(n)} \right) \right|}{\lambda_1 \ldots \lambda_n} = \frac{D}{\lambda_1 \ldots \lambda_n} ,$$

since $b^{(1)}, \ldots, b^{(n)}$ is a basis for the lattice. We have, therefore,

$$V \geq \frac{2^n D}{n! \, \lambda_1 \ldots \lambda_n} , \quad \text{or} \quad \frac{2^n}{n!} \leq \frac{V}{D} \cdot \lambda_1 \ldots \lambda_n ,$$

which is the left side of the inequality in (4). The right side of that inequality will be an immediate consequence of the following

Lemma 2. *Let* $c^{(1)}, \ldots, c^{(n)}$ *be any set of linearly independent vectors in* $\wedge$, *such that the sequence*

$$\nu_1 = f(c^{(1)}), \quad \ldots, \quad \nu_n = f(c^{(n)})$$

is a non-decreasing sequence. Then

$$\lambda_k \leq \left(\frac{3}{2} \right)^{k-1} \nu_k , \quad k = 1, \ldots, n .$$

Consider the vectors $b^{(1)}, \ldots, b^{(k-1)}$, and the vectors $c^{(1)}, \ldots, c^{(k)}$, where $2 \leq k \leq n$. At least one of the latter vectors is linearly independent of $b^{(1)}, \ldots, b^{(k-1)}$. Call it $c^{(h)}$, $h \leq k$. Now consider the set

$$b^{(1)}, \ldots, b^{(k-1)}, c^{(h)} .$$

Since $b^{(1)}, \ldots, b^{(k-1)}$ is a primitive set of vectors, we can construct a vector b in $\wedge$ of the form

$$b = \rho_1 b^{(1)} + \ldots + \rho_{k-1} b^{(k-1)} + \rho_k c^{(h)} ,$$

such that

$$b^{(1)}, \ldots, b^{(k-1)}, b$$

is a primitive set. Note that ρ_k^{-1} is an *integer*, and that we may assume that

$$-\frac{1}{2} \le \rho_j < \frac{1}{2}, \quad \text{for } j = 1, \ldots, k-1 \,.$$

Now from the extremal property of the basis vectors $b^{(1)}, \ldots, b^{(k-1)}$, it follows that

$$f(b) \ge \lambda_k \,,$$

but by the convexity property of the gauge function [Lecture I, Theorem 6], we have

$$f(b) \le |\rho_1|\lambda_1 + \ldots + |\rho_{k-1}|\lambda_{k-1} + |\rho_k|\nu_h \le \frac{(\lambda_1 + \ldots + \lambda_{k-1})}{2} + \nu_k \,;$$

therefore

$$\lambda_k \le \frac{(\lambda_1 + \ldots + \lambda_{k-1})}{2} + \nu_k \,.$$

Suppose now that

$$\lambda_j \le \left(\frac{3}{2}\right)^{j-1} \nu_j \,, \quad \text{for } j = 1, \ldots, k-1 \,.$$

Then by substituting in the inequality just proved, we have

$$\lambda_k \le \frac{\left[\left(\frac{3}{2}\right)^0 + \left(\frac{3}{2}\right)^1 + \ldots + \left(\frac{3}{2}\right)^{k-2}\right]}{2}\nu_k + \nu_k$$

(5)

$$= \left[\left(\frac{3}{2}\right)^{k-1} - 1 + 1\right]\nu_k = \left(\frac{3}{2}\right)^{k-1}\nu_k \,.$$

Since $\lambda_1 \le \nu_1$, the lemma follows by induction.

The proof of the First Finiteness Theorem follows immediately. We have

$$\lambda_1 \ldots \lambda_n \le \mu_1 \ldots \mu_n \left(\frac{3}{2}\right)^{n(n-1)/2} \,,$$

which completes the proof.

The best possible constant on the right-hand side is still (?) unknown. The result given is due to Mahler, and to Weyl, in papers published in 1938, and in 1942.

Note that if the vectors $c^{(1)}, \ldots, c^{(n)}$ also form a basis for $\wedge$, then by the use of the lemma, we can prove that

$$\nu_k \le \left(\frac{3}{2}\right)^{k-1}\lambda_k \,, \quad k = 1, \ldots, n \,.$$

Phillips Memorial
Library
Providence College

§7. Criteria for reduction

The method given for finding a reduced basis for a lattice is awkward to apply, because we must work with sets of primitive vectors. It would be convenient to have a test for deciding whether a set of vectors is primitive. Such a test is given by

Lemma 3. *Let $b^{(1)}, \ldots, b^{(n)}$ be a basis for the lattice $\wedge$ in $\mathbb{R}^n$. Let any vector b in $\wedge$ be written as*

$$g_1 b^{(1)} + \ldots + g_n b^{(n)} ,$$

where $g_1, \ldots, g_n$ are integers. Then the set

$$b^{(1)}, \ldots, b^{(k-1)}, b$$

is primitive if and only if the greatest common divisor of $g_k, g_{k+1}, \ldots, g_n$ is 1.

Suppose that the set *is* primitive. Then it can be completed to a basis for $\wedge$. The transformation from the original basis to the new basis may be written as follows:

$$
\begin{aligned}
b^{(1)} &= & b^{(1)} \\
&\vdots& \\
(6) \qquad b^{(k-1)} &= & b^{(k-1)} \\
b &= & g_1 b^{(1)} + \ldots + g_n b^{(n)} \\
&\vdots&
\end{aligned}
$$

The matrix of the transformation is

$$
(7) \qquad
\begin{pmatrix}
1 & & & & g_1 & \cdots \\
& \ddots & 0 & & \vdots & \\
0 & & \ddots & & & \\
& & & 1 & g_{k-1} & \cdots \\
0 & \cdots & 0 & 0 & g_k & \cdots \\
\vdots & & & \vdots & \vdots & \\
0 & \cdots & & 0 & g_n & \cdots
\end{pmatrix}.
$$

It must be unimodular, since it transforms one basis for $\wedge$ to another. The determinant of the matrix reduces to

$$
(8) \qquad \det
\begin{pmatrix}
g_k & \cdots \\
g_{k+1} & \cdots \\
\vdots & \\
g_n & \cdots
\end{pmatrix}.
$$

Since the absolute value of the determinant is 1, and since all the elements are integers it follows that the greatest common divisor (g.c.d.) of the elements in the first column must be 1, for otherwise the determinant would have a factor larger than 1.

Conversely if the greatest common divisor is 1, by the theorem of Gauss [Lecture VIII, Theorem 32], the $n - k + 1$ rowed submatrix in (8) can always be completed to a unimodular matrix, and so the matrix in (7) is unimodular. This proves that

$$b^{(1)}, \ldots, b^{(k-1)}, b$$

is part of a basis for $\wedge$, and therefore it must be a primitive set of vectors.

We may now write the extremal property of the $b^{(k)}$, $k = 1, \ldots, n$, as follows:

(9) $$f\left(g_1 b^{(1)} + \ldots + g_n b^{(n)}\right) \geq f\left(b^{(k)}\right) = \lambda_k ,$$

for all integral values of $g_1, \ldots, g_n$, such that

(10) $$\text{g.c.d.}\{g_k, g_{k+1}, \ldots, g_n\} = 1 .$$

The basis vectors $b^{(k)}$, $k = 1, \ldots, n$, which satisfy (9) and (10) are said to constitute a *reduced basis*. Denote by $\mathcal{F}$ the space of matrices B whose columns form a reduced basis.

The preceding arguments have shown that starting with any matrix C in the space $\mathfrak{A}$, we can always find a unimodular matrix U, such that

$$CU = B ,$$

and B lies in $\mathcal{F}$. The points C and B are called *right equivalent*. If the reduced basis were unique for every point in $\mathfrak{A}$, there would be one and only one point in $\mathcal{F}$ equivalent to it, and our problem would be solved.

§8. Use of a quadratic gauge function

The general case of an arbitrary gauge function has not been developed further. We shall now restrict ourselves to the case where

$$f(x) = |x| = \left(x_1^2 + \ldots + x_n^2\right)^{1/2} , \quad x = (x_1, \ldots, x_n) ,$$

the Euclidean length of the vector x. We state again the procedure to be followed in this case to obtain a reduced basis:

Let $b^{(1)}, \ldots, b^{(k-1)}$ be the first $k - 1$ vectors of the reduced basis. Then among all vectors b, such that

$$b^{(1)}, \ldots, b^{(k-1)}, b$$

form a primitive set, $b^{(k)}$ is a vector of shortest length.

Another way of stating it is the following:

The basis $b^{(1)}, \ldots, b^{(n)}$ is a reduced basis, if the length of any vector of the form

(11) $$g_1 b^{(1)} + \ldots + g_n b^{(n)} ,$$

where $g_1, \ldots, g_n$ are integers, is greater than or equal to the length of $b^{(k)}$, whenever

(12)   g.c.d.$\{g_k, g_{k+1}, \ldots, g_n\} = 1$.

Clearly (11) and (12) give rise to an infinite set of conditions. We shall show that they are equivalent to a finite number of inequalities. This will be the *Second Finiteness Theorem*.

Before proving it, we will show that the problem of reducing a lattice with the gauge function $f(x) = |x|$ is the same as the problem of reducing an arbitrary positive-definite quadratic form.

Let $a^{(1)}, \ldots, a^{(n)}$ be an arbitrary basis for a lattice Λ in $\mathbb{R}^n$. Let A represent the matrix whose columns are the vectors $a^{(1)}, \ldots, a^{(n)}$. Consider the linear transformation given by

$$x = \sum_{k=1}^{n} a^{(k)} y_k = Ay .$$

We have

$$|x|^2 = \sum_{k,j=1}^{n} a^{(k)} \cdot a^{(j)} \cdot y_k y_j = \sum_{k,j=1}^{n} s_{kj} \cdot y_k y_j ,$$

where

$$s_{kj} = a^{(k)} \cdot a^{(j)} = s_{jk} .$$

Note that s_{kj} is the scalar product of the vectors $a^{(k)}$ and $a^{(j)}$.

The quadratic form

$$\sum_{k,j=1}^{n} s_{kj} y_k y_j$$

is positive-definite, since $\det A \neq 0$ and $|x|^2 > 0$, unless $x = 0$. Let S be the matrix whose elements are s_{kj}. *We write $S > 0$ to indicate that S is positive-definite.* Note that $S = A'A$, where A' is the transpose of A [nowadays one writes ${}^t A$].

We have just seen that if $S = A'A$, then $S > 0$. Conversely if $S' = S$ and $S > 0$, there exists a non-singular matrix A, such that $S = A'A$. This follows from a consideration of the quadratic form $\sum_{k,j=1}^{n} s_{kj} y_k y_j$. By the method of completing a square, we can write the quadratic form as the sum of squares

$$\sum_{j=1}^{n} \left(a_j^{(1)} y_1 + \ldots + a_j^{(n)} y_n \right)^2 .$$

Let A be the matrix whose elements are $a_j^{(k)}$. Then $S = A'A$. Suppose B is another matrix such that $S = B'B$. Then we have

$$E = B'^{-1} E B' = B'^{-1} S S^{-1} B' = B'^{-1} A' A B^{-1} B'^{-1} B'$$
$$= B'^{-1} A' A B^{-1} = (AB^{-1})' (AB^{-1}) ,$$

so that if we put $O = AB^{-1}$, we have $O'O = E$. This shows that O is an orthogonal matrix, and that $A = OB$. Therefore the matrix A is determined up to an orthogonal matrix. In other words, the basis is determined except for a rotation and perhaps a reflection.

Suppose A is replaced by AU, where U is unimodular, then S will be replaced by $U'SU$, that is, the variables in the quadratic form have been subjected to a unimodular transformation. We write

$$U'SU = S[U] .$$

The quadratic forms whose matrices are S and $S[U]$ are called *equivalent*. [See § 2 of Lecture IX.]

§9. Reduction of positive-definite quadratic forms

Suppose that the basis $a^{(1)}, \ldots, a^{(n)}$ is reduced. We then say that the *matrix* $S = A'A$ is *reduced* [see (14)]. We have

$$f(x) = f(Ay) = \left[\sum_{k,j=1}^{n} s_{kj} y_k y_j \right]^{1/2} = [Q(y)]^{1/2} , \quad \text{say} .$$

The reduction conditions (11) and (12) become the following:

(13) $$Q(g_1, \ldots, g_n) \geq s_{kk} = s_k , \quad \text{say} ,$$

for all integers $g_1, \ldots, g_n$ such that the greatest common divisor of $g_k, g_{k+1}, \ldots, g_n$ is 1.

If we multiply every basis vector $a^{(k)}$ by -1, the criteria for reduction will not be changed. Consider

$$s_{1k} = a^{(1)} \cdot a^{(k)} .$$

If we change the sign of $a^{(k)}$ when $k > 1$, we will replace s_{1k} by $-s_{1k}$. We can therefore arrange it so that

(14) $$s_{1k} \geq 0 , \quad k > 1 .$$

This fixes the signs of the $a^{(k)}$ except for replacing A by $-A$, which does not change S.

Conditions (13) and (14) are the reduction conditions for a positive-definite quadratic form $Q(y)$.

Lecture XI

§1. Space of symmetric matrices

Let S be a symmetric matrix of n rows and n columns, and denote its elements by s_{ij}. Because of the symmetry, the matrix has only $n(n+1)/2$ independent elements. Consider those elements as the rectangular coordinates of a point in a space S of $n(n+1)/2$ dimensions. The space S will be called the *space of symmetric matrices*.

Those symmetric matrices which are positive-definite will form a subspace $\mathcal{P}$ of the space S. In other words, $\mathcal{P}$ will be the space of all symmetric matrices S, such that the quadratic form

$$Q(x) = \sum_{i,j=1}^{n} s_{ij} x_i x_j > 0 ,$$

for all $x \neq 0$. We shall prove the following:

$\mathcal{P}$ *is an open convex cone with centre at the origin.* That $\mathcal{P}$ is a cone with centre at the origin is trivial, because if S belongs to $\mathcal{P}$, then λS will also be positive-definite for any real positive λ, and so λS will belong to $\mathcal{P}$.

Suppose now that S and T belong to $\mathcal{P}$. We show that $\lambda S + \mu T$ is also positive-definite for $\lambda > 0$, $\mu > 0$, so that $\mathcal{P}$ is convex. The positive-definiteness follows from the fact that

$$\sum_{i,j=1}^{n} (\lambda s_{ij} + \mu t_{ij}) x_i x_j = \lambda \sum_{i,j=1}^{n} s_{ij} x_i x_j + \mu \sum_{i,j=1}^{n} t_{ij} x_i x_j > 0 ,$$

unless $x_i = 0$ for all i.

When does S belong to $\mathcal{P}$? We shall prove the well-known fact that S *belongs to $\mathcal{P}$ if and only if all the principal minors of S are positive.* The first principal minor is s_{11}; the second is the minor whose diagonal elements are s_{11}, s_{22}; the third has the diagonal elements s_{11}, s_{22}, s_{33}; and so on.

Let

$$Q(x) = \sum_{i,j=1}^{n} s_{ij} x_i x_j$$

be the quadratic form associated to S. If we complete the square in $Q(x)$, first using all terms involving x_1, then all terms involving x_2, and so on, we find

that
$$Q(x) = c_1 \left(x_1 + c_{12}x_2 + \ldots + c_{1n}x_n\right)^2$$
$$+ c_2 \left(x_2 + c_{23}x_3 + \ldots + c_{2n}x_n\right)^2 + \ldots + c_n x_n^2 \ .$$

Now if $c_1, \ldots, c_n$ are positive, S is positive-definite; conversely, if S is positive-definite, then $c_1, \ldots, c_n$ are positive. Let $x_{k+1} = \ldots = x_n = 0$, then we have the identity

$$\sum_{i,j=1}^{k} s_{ij}x_ix_j = c_1(x_1 + c_{12}x_2 + \ldots + c_{1k}x_k)^2$$

$$+ c_2(x_2 + c_{23}x_3 + \ldots + c_{2k}x_k)^2 + \ldots + c_k x_k^2 \ .$$

If
$$S_{(k)} = \begin{pmatrix} s_{11} & \cdots & s_{1k} \\ \vdots & & \vdots \\ s_{k1} & \cdots & s_{kk} \end{pmatrix},$$

$$x_{(k)} = \begin{pmatrix} x_1 \\ \vdots \\ x_k \end{pmatrix},$$

and
$$A_{(k)} = \begin{pmatrix} \sqrt{c_1} & \sqrt{c_1}c_{12} & \cdots & \sqrt{c_1}c_{1k} \\ 0 & \sqrt{c_2} & \cdots & \sqrt{c_2}c_{2k} \\ \vdots & & \ddots & \vdots \\ 0 & \cdots & 0 & \sqrt{c_k} \end{pmatrix},$$

then the above identity amounts to

$$x'_{(k)}S_{(k)}x_{(k)} = x'_{(k)}A'_{(k)}A_{(k)}x_{(k)} \ .$$

If we express the fact that the determinants of the matrices must be equal, we find that the principal minor of S of order k equals $c_1 \ldots c_k$. This proves the theorem that S is positive-definite if and only if its principal minors are positive. It follows that $\mathcal{P}$ is an open subset of $\mathcal{S}$.

§2. Reduction of positive-definite quadratic forms

Let A be a non-singular matrix with real elements. If we apply the transformation with matrix A to the quadratic form $Q(x)$, corresponding to a matrix S in $\mathcal{P}$, we get a new quadratic form whose matrix is $A'SA$. We shall write this matrix as $S[A]$.

The passage from S to $S[A]$ will be a mapping of $\mathcal{P}$ onto itself, since $S > 0$ and $\det A \neq 0$ imply that $A'SA > 0$. Instead of considering general A, we shall restrict ourselves to the case where A is a unimodular matrix U.

Out of the space of matrices equivalent to S, that is, matrices which are representable as $U'SU$, we wish to find one representative. In the previous lectures we proved that we can always find a reduced matrix T equivalent to the positive-definite matrix S.

A positive-definite matrix T, or the corresponding quadratic form $Q(x) = x'Tx = T[x]$, is said to be *reduced*, if

$$(1) \qquad\qquad Q(g) \geq t_{kk} ,$$

for all integral vectors g, such that

$$(2) \qquad\qquad \text{g.c.d.} \{g_k, \ldots, g_n\} = 1 , \quad k = 1, \ldots, n ,$$

and if, in addition,

$$(3) \qquad\qquad t_{1j} \geq 0 , \quad j = 2, \ldots, n .$$

For convenience we shall usually suppress one index for the diagonal elements of a symmetric matrix, so that, for instance we write t_k instead of t_{kk}.

The First Finiteness Theorem proved in the last lecture now becomes

$$\frac{4^n}{(n!)^2} \leq \frac{V^2}{\det T} \cdot t_1 \ldots t_n \leq \left[2^n \left(\frac{3}{2} \right)^{\frac{n(n-1)}{2}} \right]^2 ,$$

where $V = $ volume of $\left\{ x \,\middle|\, \sqrt{x_1^2 + \ldots + x_n^2} < 1 \right\} = \frac{\pi^{n/2}}{\Gamma(\frac{n}{2}+1)}$. Substituting for V, we get the following inequality:

$$(4) \qquad \left(\frac{4}{\pi} \right)^n \frac{\left[\Gamma \left(\frac{n}{2} + 1 \right) \right]^2}{\left[\Gamma(n+1) \right]^2} \leq \frac{t_1 \ldots t_n}{\det T} \leq \left(\frac{4}{\pi} \right)^n \left(\frac{3}{2} \right)^{n(n-1)} \left[\Gamma \left(\frac{n}{2} + 1 \right) \right]^2 .$$

The left-hand side of this inequality may be replaced by 1, and this is the best possible result. In other words, if $S' = S > 0$, then

$$(5) \qquad\qquad s_1 \ldots s_n \geq \det S .$$

This is essentially equivalent to a theorem of Hadamard's. The proof follows from the decomposition given above. We showed that $S = A'A$, where A is the matrix

$$(6) \qquad \begin{pmatrix} \sqrt{c_1} & \sqrt{c_1} c_{12} & \cdots & \sqrt{c_1} c_{1n} \\ 0 & \sqrt{c_2} & \cdots & \sqrt{c_2} c_{2n} \\ \vdots & \ddots & \ddots & \vdots \\ 0 & \cdots & 0 & \sqrt{c_n} \end{pmatrix} .$$

Let $c^{(i)}$ represent the ith column vector of A, then

$$\det A = \sqrt{c_1} \ldots \sqrt{c_n} \leq \left| c^{(1)} \right| \ldots \left| c^{(n)} \right| ,$$

and

$$\det S = (\det A)^2 \leq \left|c^{(1)}\right|^2 \cdots \left|c^{(n)}\right|^2 = s_1 \ldots s_n \, ,$$

since s_i is the scalar product of $c^{(i)}$ by itself.

The best possible value of the right-hand side is not known. We shall determine the best possible value for $n = 2$ and $n = 3$.

§3. Consequences of the reduction conditions

It is clear that the reduction conditions (1) and (2) imply that

(7)
$$t_1 \leq t_2 \leq \ldots \leq t_n \, ,$$

since t_k is a minimum value defined under successively stronger conditions.

We shall derive an additional consequence of conditions (1) and (2). Let i, j be integers, such that $1 \leq i < j \leq n$, $n \geq 2$. Take $\pm g_i = g_j = 1$, all other g_k's being zero. Then condition (2) is satisfied for $k = j$, and therefore

$$Q(0, \ldots, 0, \pm 1, 0, \ldots, 0, 1, 0, \ldots, 0) \geq t_j \, ,$$

that is

$$t_i \pm 2t_{ij} + t_j \geq t_j \, ,$$

or

(8)
$$|2t_{ij}| \leq t_i \, , \quad i < j \, .$$

We may describe this in words by the statement that all elements to the right of the diagonal in any row are not greater in absolute value than one-half of the diagonal element. Using condition (3) we may finally write

(9)
$$\begin{cases} 0 \leq 2t_{1j} \leq t_1 \, , & j > 1 \, , \\ |2t_{ij}| \leq t_i \, , & i < j \, , \\ t_1 \leq t_2 \leq \ldots \leq t_n \, . \end{cases}$$

§4. The case n=2

Let S be the matrix $\begin{pmatrix} a & b \\ b & c \end{pmatrix}$. For S to be positive-definite, we must have

(10)
$$a > 0 \, , \quad ac - b^2 = \Delta > 0 \, .$$

This will be assumed. Put $Q(x, y) = ax^2 + 2bxy + cy^2$.

We shall now apply the reduction theory to S. Conditions (1) and (2) imply that

(11)
$$Q(x, y) \geq a \, ,$$

as long as g.c.d.$\{x, y\} = 1$. It is obvious that (11) will hold if x, y are not both zero.

We also have

(12) $$Q(x,y) \geq c \, ,$$

if $y = \pm 1$.

From conditions (9) we deduce that

(13) $$a \leq c \, , \quad 0 \leq 2b \leq a \, .$$

We shall show that if (13) is satisfied, the infinite number of conditions implied by (11) and (12) are all satisfied, so that *condition* (13) *is a necessary and sufficient condition for reduction.*

The proof is simple. Using condition (13), we have

$$Q(x,y) = ax^2 + 2bxy + cy^2$$
$$\geq ax^2 - a|xy| + ay^2 + (c-a)y^2$$
$$= a(x^2 - |xy| + y^2) + (c-a)y^2 \, .$$

Now $x^2 - |xy| + y^2$ is a positive-definite form with integral coefficients, and therefore it is equal to or greater than 1, unless $(x,y) = (0,0)$. Since $c - a$ is non-negative, we have

$$Q(x,y) \geq a \, ,$$

and if $y \neq 0$, we have

$$Q(x,y) \geq a + (c-a) = c \, .$$

This proves that (13) implies (11) and (12).

The First Finiteness Theorem implies (4), which when applied to this gives

$$\frac{ac}{ac - b^2} \leq \left(\frac{4}{\pi}\right)^2 \left(\frac{3}{2}\right)^2 = \frac{36}{\pi^2} = 3.64756\ldots \, .$$

[Note also that (13) together with the condition $a > 0$ already implies that $Q(x,y)$ is positive-definite.] We shall show that the exact bound for the right-hand side is $\frac{4}{3}$, so that

(14) $$\frac{ac - b^2}{ac} \geq \frac{3}{4} \, .$$

The proof is a consequence of conditions (13). We have

$$\frac{ac - b^2}{ac} \geq \frac{ac - \frac{a^2}{4}}{ac} \geq 1 - \frac{a}{4c} \geq 1 - \frac{1}{4} = \frac{3}{4} \, .$$

We may write (14) as follows:

$$ac \leq \frac{4}{3} \left(ac - b^2 \right) \, ,$$

and since $a \leq c$, we have the result proved before [Lecture IX, § 5] that

(15) $$a \leq 2\sqrt{\frac{\Delta}{3}} \, ,$$

where a is the minimum of the positive-definite quadratic form $Q(x,y)$.

§5. Reduction of lattices of rank two

We have shown that every positive-definite binary quadratic form is equivalent to a reduced form $ax^2 + 2bxy + cy^2$, where

$$0 < a \leq c, \quad 0 \leq 2b \leq a.$$

Because of the connection between positive-definite quadratic forms and lattices, we shall obtain the conditions for a reduced basis.

Suppose the column vectors $a^{(1)}, a^{(2)}$ of the matrix A, form a basis for a lattice of rank two. Put

$$A'A = S = \begin{pmatrix} a & b \\ b & c \end{pmatrix}.$$

Note that

$$a = \left| a^{(1)} \right|^2, \quad c = \left| a^{(2)} \right|^2, \quad b = a^{(1)} \cdot a^{(2)}.$$

If ω denotes the angle between the vectors $a^{(1)}$ and $a^{(2)}$, then

$$\cos \omega = \frac{a^{(1)} \cdot a^{(2)}}{\left| a^{(1)} \right| \cdot \left| a^{(2)} \right|} = \frac{b}{\sqrt{ac}}.$$

The reduction conditions imply that

$$\sqrt{ac} \geq a \geq 2b \geq 0, \quad \text{or} \quad \frac{1}{2} \geq \cos \omega \geq 0,$$

and we must have

(16) $$60° \leq \omega \leq 90°.$$

This proves that *in a given lattice of rank two, there exists a basis such that the angle between the basis vectors is between 60 and 90 degrees.*

By means of this fact, we can prove a result equivalent to (15):

In any lattice of rank two with determinant D, there exists a vector v, *not* the origin, such that

(17) $$|v|^2 \leq \frac{2D}{\sqrt{3}}.$$

It is clear that the area of the parallelogram formed by any basis vectors is D. We then have

$$\left| a^{(1)} \right| \cdot \left| a^{(2)} \right| \cdot \sin \omega = D,$$

or $\sqrt{ac} = \frac{D}{\sin \omega}$. It follows that

$$\left| a^{(1)} \right|^2 = a \leq \sqrt{ac} = \frac{D}{\sin \omega} \leq \frac{2D}{\sqrt{3}},$$

since $60° \leq \omega \leq 90°$, and therefore the vector $v = a^{(1)}$ satisfies (17).

§6. The case $n=3$

We shall again determine a finite set of reduction conditions, and we shall also find the best possible constant in the First Finiteness Theorem.

Put

$$Q(x,y,z) = ax^2 + by^2 + cz^2 + 2dxy + 2exz + 2fyz .$$

We assume that $Q(x,y,z)$ is positive-definite. In particular, we then have $a > 0$.

Conditions (9) state that

$$
\begin{array}{lll}
\text{(i)} & 0 \le 2d \le a , & 0 \le 2e \le a , \\
\text{(ii)} & -b \le 2f \le b , & \\
\text{(iii)} & a \le b \le c .
\end{array}
$$

(18)

These eight conditions are not sufficient. We need another condition, which we obtain by considering $Q(-1,1,1)$. From conditions (1) and (2) we find that

$$Q(-1,1,1) = a + b + c - 2d - 2e + 2f \ge c ,$$

or

(19) $-2f \le a + b - 2(d+e) .$

If $f \ge 0$, this condition is a consequence of (i) and (iii) of (18), because the right-hand side in (19) is always non-negative since

$$2d + 2e \le a + a \le a + b .$$

If $f < 0$, however, condition (19) is independent of the conditions in (18).

Put $f\theta$ for f, where $\theta = \pm 1$, $f \ge 0$, so that

$$Q(x,y,z) = ax^2 + by^2 + cz^2 + 2dxy + 2exz + 2f\theta yz .$$

Then the conditions on f may be written as follows:

$$-2f\theta \le a + b - 2(d+e) ,$$
$$0 \le 2f \le b .$$

(20)

We shall prove that conditions (20) together with the following:

$$0 \le 2d \le a , \qquad 0 \le 2e \le a ,$$
$$a \le b \le c ,$$

(21)

are sufficient to assure that $Q(x,y,z)$ be reduced. We must prove that conditions (1) and (2) are satisfied, that is to say

$$
\begin{array}{ll}
Q(x,y,z) \ge a , & \text{if g.c.d.}\{x,y,z\} = 1 ; \\
Q(x,y,z) \ge b , & \text{if g.c.d.}\{y,z\} = 1 ; \\
Q(x,y,z) \ge c , & \text{if } z = \pm 1 .
\end{array}
$$

We shall prove the third inequality not only for $z = \pm 1$, but for any $z \neq 0$. Since c is not smaller than b, it then follows that the second inequality need be proved only for $z = 0$, $y = \pm 1$. We shall prove the second inequality for any $y \neq 0$. Since b is not smaller than a, it then follows that the first inequality need be proved only for $y = z = 0$ and $x = \pm 1$. However

$$Q(x, 0, 0) = ax^2 \geq a, \quad \text{for } x \neq 0.$$

The second case, where

$$Q(x, y, 0) = ax^2 + by^2 + 2dxy$$

has been treated before in the discussion of the positive-definite quadratic form in *two* variables. Since $0 < a \leq b$, $0 \leq 2d \leq a$, $Q(x, y, 0)$ is a reduced, positive-definite quadratic form of two variables, and therefore

$$Q(x, y, 0) \geq b, \quad \text{if } y \neq 0.$$

Let us now consider the case of the third inequality. If $x = 0$, then

$$Q(0, y, \theta z) = by^2 + cz^2 + 2fyz,$$

and just as before, this is a reduced, positive-definite quadratic form in two variables, so that $Q(0, y, \theta z) \geq c$, if $z \neq 0$, and therefore

$$Q(0, y, z) \geq c, \quad \text{if } z \neq 0.$$

A similar argument can be used when $y = 0$.

We may therefore assume that x, y, z are all *different from zero*. We write

$$(22) \quad \begin{aligned} Q(x, y, z) &= (a - d - e)x^2 + (b - d - f)y^2 \\ &\quad + (c - e - f)z^2 + d(x + y)^2 + e(x + z)^2 + f(y + \theta z)^2. \end{aligned}$$

From conditions (20), (21) it follows that all the coefficients are non-negative. We shall prove that $Q(x, y, z) \geq c$, if none of the variables x, y, z is zero.

If at least two of the three terms $x + y$, $x + z$, $y + \theta z$ are different from zero, then the sum of the last three summands in (22) is greater than or equal to

$$(23) \quad d + e, \quad \text{or } e + f, \quad \text{or } d + f,$$

depending on which two terms are different from zero.

If exactly two of the three terms are zero, then the third term must be a multiple of 2. For example, if $x + y = x + z = 0$, then

$$0 = y - z \equiv y + z \pmod{2}.$$

A similar argument holds for the other cases. We conclude that the sum of the last three summands in (22) will now be greater than or equal to

(24) $2d$,   or $2e$,   or $2f$,

depending on which term is different from zero. Suppose now that d is the smallest of the three numbers d, e, f. Then combining (22) with (23) or (24), we get

$$Q(x, y, z) \geq a + b + c - 2d - 2e - 2f + 2d$$
$$= (a - 2e) + (b - 2f) + c \geq c ,$$

since the terms in brackets are non-negative, by conditions (20) and (21).

If the last three terms are all zero, that is, if

$$x + y = x + z = y + \theta z = 0 ,$$

then $y = z$, $\theta = -1$, and we have

$$Q(x, y, z) \geq a + b + c - 2d - 2e - 2f .$$

But the first inequality of condition (20) states that

$$a + b - 2d - 2e - 2f \geq 0 .$$

Therefore $Q(x, y, z) \geq c$, in this case also.

Note that again the infinite set of original reduction conditions has been replaced by only nine conditions – those given by (20) and (21). This is another illustration of the *Second Finiteness Theorem*, which we shall discuss in the next lecture. It states that, in all cases, the infinite set of reduction conditions (1) and (2) may be replaced by a finite set.

We shall now find the exact bound for the right-hand side of the inequality in the First Finiteness Theorem. We prove that

(25) $abc \leq 2\Delta$.

This result was first discovered by Gauss.

We shall establish (25) by trying to minimize $\frac{\Delta}{abc}$. We have

$$\Delta = \det \begin{pmatrix} a & d & e \\ d & b & f\theta \\ e & f\theta & c \end{pmatrix} = abc + 2def\theta - af^2 - be^2 - cd^2 .$$

Suppose first that $\theta = -1$. Assuming that $f \neq 0$, we see that Δ is a decreasing function of f, so that Δ is made smaller if we take f as large as permissible under the conditions in (20), that is

$$f = \min \left\{ \frac{1}{2}(a + b - 2(d + e)), \frac{b}{2} \right\} .$$

[We shall deal with the case $f = 0$ when we consider the alternative $\theta = +1$.]

If $f = \frac{1}{2}(a + b - 2(d + e))$, then $a - 2d \leq 2e$, and we have

$$\Delta = abc - de\,(a + b - 2(d + e)) - \frac{a}{4}\,(a + b - 2(d + e))^2 - be^2 - cd^2$$

$$= -e^2\,((-2d + a) + b) + \{\text{a term which is linear in } e, \text{ and}$$
$$\text{a term which does not depend on } e\}\,.$$

Taking into account conditions (21), we see that this equation represents a parabola in the (e, Δ)-plane, which opens towards the negative Δ-axis. Because of (21), e is subject to the condition $0 \leq 2e \leq a$. The minimal value of Δ is attained therefore either at $e = 0$ or at $e = \frac{a}{2}$.

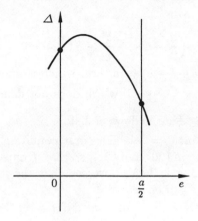

If $e = 0$, then by (21), $0 \leq a - 2d$, and since $a - 2d \leq 2e = 0$, by assumption (in this case), we have $2d = a$, and $f = \frac{b}{2}$, which will be treated separately.
If $e = \frac{a}{2}$, we get

$$\Delta = abc - \frac{ad}{2}(b - 2d) - \frac{a}{4}(b - 2d)^2 - \frac{a^2 b}{4} - cd^2$$

$$= -d^2(-a + a + c) + \{\text{a term which is linear in } d, \text{ and a term}$$
$$\text{which does not depend on } d\}\,.$$

By (21), d satisfies the condition $0 \leq 2d \leq a$, and by the same argument as above, we shall have $d = 0$, or $d = \frac{a}{2}$. Now if $d = 0$, we get

(26)
$$\Delta = abc - \frac{ab^2}{4} - \frac{a^2 b}{4} = abc\left(1 - \frac{b}{4c} - \frac{a}{4c}\right)$$
$$\geq abc\left(1 - \frac{1}{4} - \frac{1}{4}\right) = \frac{1}{2}abc\,,$$

since $0 < a \leq b \leq c$.

If, on the other hand, $d = \frac{a}{2}$, we get

$$\Delta = abc - \frac{a^2}{4}(b-a) - \frac{a}{4}(b-a)^2 - \frac{a^2 b}{4} - \frac{a^2 c}{4}$$

$$= abc - \frac{ab^2}{4} - \frac{a^2 c}{4} \geq \frac{1}{2}abc, \qquad \text{as in (26)}.$$

Returning to the possibility $f = \frac{b}{2}$, we see that then $2e \leq a - 2d$, and

$$\Delta = abc - bde - \frac{ab^2}{4} - be^2 - cd^2 .$$

Since the derivative $\frac{\partial \Delta}{\partial e} \leq 0$, we shall choose the maximum possible value for e, consistently with the conditions $2e \leq a$, and $2e \leq a - 2d$, namely $e = \frac{a}{2} - d$, so that we get

$$\Delta = abc - bd\left(\frac{a}{2} - d\right) - \frac{ab^2}{4} - b\left(\frac{a}{2} - d\right)^2 - cd^2$$
$$= -d^2(-b + b + c) + \{\text{a term which is linear in } d, \text{ and a term}$$
$$\text{which does not depend on } d\} .$$

The conclusion $\Delta \geq \frac{1}{2}abc$, follows as before.

Suppose now that $\theta = +1$. The first condition in (20) then follows from the other conditions in (20) and (21), so that f must only satisfy the condition $0 \leq 2f \leq b$. Arguing as before, we see that the minimal value of Δ will be attained either at $f = 0$ or at $f = \frac{b}{2}$.

If $f = 0$, then $\Delta = abc - be^2 - cd^2$. If $f = \frac{b}{2}$, then

$$\Delta = abc + bde - \frac{ab^2}{4} - be^2 - cd^2 = abc - b\left(\frac{ab}{4} - de\right) - be^2 - cd^2 .$$

Because of (21), we have $\frac{ab}{4} - de \geq 0$. It follows that the value of Δ at $f = \frac{b}{2}$ is at least as low as the value of Δ at $f = 0$. Hence

$$\Delta \geq abc + bde - \frac{ab^2}{4} - be^2 - cd^2 .$$

Considering the right-hand side of this inequality as a function of e, and arguing as before, we conclude that the minimum of the right-hand side is attained for $e = \frac{a}{2}$, so that

$$\Delta \geq abc + \frac{abd}{2} - \frac{ab^2}{4} - \frac{a^2 b}{4} - cd^2 .$$

The right-hand side here, now considered as a function of d, has its minimum at $d = \frac{a}{2}$, so that

$$\Delta \geq abc + \frac{a^2 b}{4} - \frac{ab^2}{4} - \frac{a^2 b}{4} - \frac{a^2 c}{4}$$

(27)

$$= abc - \frac{ab^2}{4} - \frac{a^2 c}{4} \geq \frac{1}{2}abc, \qquad \text{as in (26)}.$$

We have therefore proved that, in all cases, $abc \leq 2\Delta$. Tracing back through the above proof, we see for which reduced forms the equality $abc = 2\Delta$ holds. It is easy to show that they are all equivalent to

$$a \left(x^2 + y^2 + z^2 + xy + xz + yz \right) \ .$$

Since a is the minimum of $Q(x, y, z)$, we have proved that

(28) $$a \leq (2\Delta)^{1/3} \ .$$

Lecture XII

§1. Extrema of positive-definite quadratic forms

Consider an arbitrary positive-definite quadratic form in n variables with determinant Δ. [By the determinant of a quadratic form is meant the determinant of the corresponding symmetric matrix.] Let r_n be the minimum value of the quadratic form on the lattice of g-points excluding the origin. In the previous lecture we showed that

$$(1) \qquad\qquad r_2 \le \sqrt{\frac{4\Delta}{3}} \ ,$$

and the equality sign holds if and only if the form is equivalent to

$$(2) \qquad\qquad r_2\left(x^2 + xy + y^2\right) \ .$$

We also showed that

$$(3) \qquad\qquad r_3 \le (2\Delta)^{1/3} \ ,$$

and the equality holds if and only if the form is equivalent to

$$(4) \qquad\qquad r_3\left(x^2 + y^2 + z^2 + xy + xz + yz\right) \ .$$

On the basis of (1) and (3) we might guess that, in general,

$$r_n \le 2\left(\frac{\Delta}{n+1}\right)^{1/n} \ ,$$

so that for $n = 4$ we would have

$$r_4 \le 2\left(\frac{\Delta}{5}\right)^{1/4} \ .$$

On the basis of (2) and (4) we would further conjecture that the equality would hold if the quadratic form is equivalent to

$$r_4\left(x^2 + y^2 + z^2 + t^2 + xy + xz + xt + yz + yt + zt\right) \ .$$

These conjectures are not true. We shall prove the following:

If r_4 is the minimum on all g-points in $\mathbb{R}^4$, excluding the origin, and Δ the determinant of a positive-definite quadratic form in 4 variables, then

$$(5) \qquad\qquad r_4 \leq (4\Delta)^{1/4} \; ,$$

and the equality sign holds if the form is equivalent to

$$(6) \qquad\qquad r_4 \left(x^2 + y^2 + z^2 + t^2 + xt + yt + zt \right) \; .$$

We shall see that this theorem is an immediate consequence of a theorem which was first proved by Mordell [J.L.M.S. *19* (1944) 3–6]. From Minkowski's Second Theorem we know that for a given n, the ratio $\frac{r_n}{\Delta^{1/n}}$ is bounded on the set of positive-definite quadratic forms in n variables. Let γ_n be the least upper bound of this ratio, for all positive-definite quadratic forms in n variables, so that in all cases we have

$$(7) \qquad\qquad r_n \leq \gamma_n \Delta^{1/n} \; .$$

Mordell's theorem states that

$$(8) \qquad\qquad \gamma_n^{n-2} \leq \gamma_{n-1}^{n-1} \; .$$

If we let $n = 4$ in (8), we get $\gamma_4^2 \leq \gamma_3^3 = 2$, so that $\gamma_4 \leq \sqrt{2} = 4^{1/4}$, but since the bound $4^{1/4}$ is attained for the form (6), we must have $\gamma_4 = 4^{1/4}$. This shows that a proof of (8) will also give a proof of (5).

We now give Mordell's *proof* of (8). It is clear that we may always restrict ourselves to the consideration of quadratic forms with determinant 1. Let $Q(x)$ be a positive-definite quadratic form in n variables with determinant 1, and let S be its matrix. We may write S as follows:

$$S = \begin{pmatrix} s_{11} & \cdots & s_{1n} \\ \vdots & & \vdots \\ s_{n1} & \cdots & s_{nn} \end{pmatrix} \; .$$

We consider also the adjoint form $Q^*(x)$, whose matrix is the adjoint of S, that is

$$S^* = \begin{pmatrix} s_{11}^* & \cdots & s_{1n}^* \\ \vdots & & \vdots \\ s_{n1}^* & \cdots & s_{nn}^* \end{pmatrix} \; ,$$

where s_{ij}^* is the cofactor [see Lecture V, § 6] of S corresponding to the element s_{ij} (since S is symmetric). Note that since $\det S$ is 1, $\det S^*$ is also 1. By a unimodular transformation we can arrange that the minimum value of $Q^*(x)$ will be attained at the point $(0, \ldots, 0, 1)$. Of course the matrix of $Q^*(x)$ will be transformed to

$$U'S^*U = T^* \; , \qquad \text{say} \; .$$

In the form $Q(x)$, which now has the matrix T, suppose that one of the variables, say x_n, is put equal to zero. Then $Q(x)$ will become a positive-definite form $Q_{n-1}(x)$, in $n-1$ variables, with the following matrix:

$$T_{n-1} = \begin{pmatrix} t_{11} & \cdots & t_{1,n-1} \\ \vdots & & \vdots \\ t_{n-1,1} & \cdots & t_{n-1,n-1} \end{pmatrix} .$$

It is clear that

(9)
$$\min_{x \in \mathbb{Z}^n - 0} Q(x) \le \min_{x \in \mathbb{Z}^{n-1} - 0} Q_{n-1}(x)$$

$[\mathbb{Z}^n - 0$ denotes the set of all g-points in $\mathbb{R}^n$ excluding the origin], since there are more variables to choose from in $Q(x)$. By the definition of the γ's, we have

(10)
$$\min_{x \in \mathbb{Z}^{n-1} - 0} Q_{n-1}(x) \le \gamma_{n-1} (\det T_{n-1})^{1/(n-1)} .$$

Since $\det T_{n-1}$ is the cofactor of t_{nn} in T, we have

(11)
$$\det T_{n-1} = t_{nn}^* .$$

Because of the unimodular transformation, the minimum value of $Q^*(x)$ is attained at $(0, \ldots, 0, 1)$, that is

(12)
$$t_{nn}^* = \min_{x \in \mathbb{Z}^n - 0} Q^*(x) \le \gamma_n (\det T^*)^{1/n} = \gamma_n ,$$

since a unimodular transformation does not change the value of the determinant.

Combining (9), (10), (11), (12), we get

$$\min_{x \in \mathbb{Z}^n - 0} Q(x) \le \gamma_{n-1} (\gamma_n)^{1/(n-1)} ,$$

but since $Q(x)$ is an arbitrary quadratic form, and since γ_n is the least upper bound of $\min_{x \in \mathbb{Z}^n - 0} Q(x)$ for all $Q(x)$, we have

$$\gamma_n \le \gamma_{n-1} (\gamma_n)^{1/(n-1)} , \quad \text{or} \quad \gamma_n^{n-2} \le \gamma_{n-1}^{n-1} .$$

This completes the proof of (8).

If we apply (8) to the case $n = 5$, we get

$$\gamma_5^3 \le 4 , \quad \text{or} \quad \gamma_5 \le 4^{1/3} .$$

It is known, however, that the exact value of γ_5 is $8^{1/5}$, so that (8) does not give the exact bound. Formula (8) will give the exact bound for $n = 8$. It is known that $\gamma_6 = \left(\frac{64}{3}\right)^{1/6}$, $\gamma_7 = (64)^{1/7}$, so that from (8) we find that

$$\gamma_8^6 \le 64 , \quad \text{or} \quad \gamma_8 \le 2 .$$

If we consider the following positive-definite form with determinant 1:

$$\sum_{i=1}^{8} x_i^2 + \left(\sum_{i=1}^{8} x_i \right)^2 - 2x_1 x_2 - 2x_2 x_8 ,$$

we find that it attains the value 2 when $x_1 = 1$, $x_i = 0$ for $1 < i \leq 8$. Since the form represents only even numbers, γ_8 is at least 2. Hence $\gamma_8 = 2$.

§2. Closest packing of (solid) spheres

Given an infinite set of equal spheres whose centres form a lattice of rank n in $\mathbb{R}^n$, what part of space is covered by these spheres? We shall consider also how the spheres may be arranged so that as much space as possible is covered. In the particular case $n = 2$, we shall be able to drop the condition that the centres form a lattice.

Let the radius of the spheres be r. Then the volume of one such sphere is $\sigma_n r^n$, where

$$(13) \qquad \sigma_n = \frac{\pi^{n/2}}{\Gamma\left(\frac{n}{2} + 1\right)} \; .$$

Suppose the centres of these spheres form a lattice. Let A be the matrix of a basis for the lattice. Then the volume of a fundamental parallelepiped is $|\det A| = D$.

In any very large cube, there will be approximately as many spheres as fundamental parallelepipeds, because to each parallelepiped we can assign one sphere. In the limit, as the length of the cube tends to infinity, the ratio of the volume covered by the spheres, to the total volume, is

$$\frac{\sigma_n r^n}{D} \; .$$

We wish to maximize this ratio. It is clear that this ratio increases with r, but in order to exclude overlapping of the spheres, r can be at most one-half the smallest distance between two different lattice points.

Any point of the lattice is specified by Ax, where x is a vector with integral coordinates. The square of the distance of that point from the origin is given by

$$x'A'Ax = x'Sx \; ,$$

if we put $S = A'A$. Note that $x'Sx$ is a positive-definite quadratic form. The distance to the nearest point from the origin will be given by the square root of the minimum of the quadratic form.

Let

$$a = \min_{x \in \mathbb{Z}^n - 0} x'Sx \; .$$

Then $2r \leq \sqrt{a}$. The largest value of the "ratio" will be obtained when $r = \frac{\sqrt{a}}{2}$; the ratio then becomes

$$(14) \qquad \frac{\sigma_n \cdot a^{n/2}}{2^n \sqrt{\Delta}} \leq \frac{\sigma_n}{2^n} \cdot \gamma_n^{n/2} = q_n \; , \quad \text{say} \; ,$$

where Δ is the determinant of S. *The best packing fraction obtainable by using spheres whose centres form a lattice is* q_n.

§3. Closest packing in two, three, or four dimensions

For $n = 2$, we have $\sigma_2 = \pi$, and $\gamma_2 = \frac{2}{\sqrt{3}}$, so that

$$q_2 = \frac{\pi}{2\sqrt{3}}.$$

The lattice corresponds to the quadratic form $x^2 + xy + y^2$.

In the previous lecture we showed that this lattice has a basis whose vectors are of equal length, and make an angle of 60°. Therefore if we assume that the centres form a lattice, the closest packing of circles will be obtained by putting one row of circles between the spaces of the lower row.

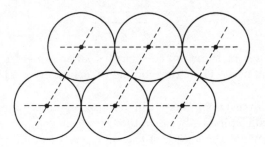

It can be proved that even if the lattice condition is dropped, the above packing is still the best.

For $n = 3$, we have $\sigma_3 = \frac{4\pi}{3}$, and $\gamma_3 = 2^{1/3}$, so that

$$q_3 = \frac{\pi}{3\sqrt{2}}.$$

The lattice corresponds to the quadratic form $x^2 + y^2 + z^2 + xy + yz + xz$.

It can be shown that all basis vectors are of equal length and make angles of 60° with each other. The centres of four adjacent spheres are located at the vertices of a regular tetrahedron. The packing is the "shot pile" arrangement, and can be described as follows.

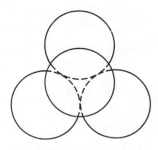

On the ground plane, the spheres are placed as the circles were in two dimensions, then in the next layer each sphere fills the hole between three adjacent spheres.

It is not known whether this arrangement will give the closest packing, if we drop the assumption that the centres form a lattice.

For $n = 4$, we have $\sigma_4 = \frac{\pi^2}{2}$, and $\gamma_4 = 4^{1/4}$, so that

$$q_4 = \frac{\pi^2}{16} \, .$$

Again it is not known whether this is the closest packing if we do not assume that the centres form a lattice. Note also that the centres do not lie at the vertices of a regular simplex.

§4. Blichfeldt's method

By using a method given by Blichfeldt we shall obtain an upper bound for the value of q_n. We prove that

(15)
$$q_n \le \left(1 + \frac{n}{2}\right) 2^{-n/2} \, .$$

This result will be obtained without assuming that the centres form a lattice.

Consider an arbitrary bounded open set $\mathcal{M}$ in $\mathbb{R}^n$ with volume V. For $t > 0$, let $V_t = t^n V$ be the volume of the set $t\mathcal{M}$. Assume there is some arrangement of spheres of radius 1, with centres at the points $x^{(1)}, x^{(2)}, \ldots,$ and that the spheres do not overlap. This implies that

$$\left| x^{(k)} - x^{(j)} \right| \ge 2 \, , \quad \text{for any } k \text{ and } j \, , \quad k \ne j \, .$$

Let β_t be the number of spheres which are completely inside the set $t\mathcal{M}$. We shall prove that

$$\limsup_{t \to \infty} \frac{\beta_t \cdot \sigma_n}{V_t} \le \left(1 + \frac{n}{2}\right) 2^{-n/2} \, .$$

The method is to assign a weight-function $x \mapsto \varphi\left(x - x^{(k)}\right)$ to each centre $x^{(k)}$, such that for any point x the sum of the weights will be ≤ 1; that is

(16)
$$\sum_{k=1}^{\infty} \varphi\left(x - x^{(k)}\right) = \psi(x) \le 1 \, .$$

Then by evaluating

(17)
$$I = \int_{t\mathcal{M}} \psi(x) \, dx$$

in two different ways, we shall get an inequality involving q_n.

As our first choice for a weight-function, we take

$$\varphi(x) = \begin{cases} 1, & |x| < 1, \\ 0, & |x| \geq 1. \end{cases}$$

Since $\left|x^{(k)} - x^{(j)}\right| \geq 2$, $k \neq j$, it is clear that all the terms in the sum will vanish except one at most, so that (16) will be satisfied. Now

(18)
$$I = \int_{tM} \psi(x)\, dx \leq V_t .$$

On the other hand, if we interchange summation and integration, we have

$$I = \sum_{k=1}^{\infty} \int_{tM} \varphi\left(x - x^{(k)}\right) dx \geq \beta_t \cdot \sigma_n ,$$

so that $V_t \geq \beta_t \cdot \sigma_n$, or $\frac{\beta_t \cdot \sigma_n}{V_t} \leq 1$, a trivial result. The reason for the trivial result is clear. We have compared the volume of tM with the volume of the spheres contained in it. Suppose, however, we take

(19)
$$\varphi(x) = \begin{cases} 1 - \frac{|x|^2}{2}, & |x| \leq \sqrt{2}, \\ 0, & |x| > \sqrt{2}. \end{cases}$$

We shall show later that (16) is satisfied.

The integral I will still be not greater than V_t, but estimating I the second way, we have

$$I = \sum_{k=1}^{\infty} \int_{tM} \varphi\left(x - x^{(k)}\right) dx \geq \alpha_t \int_{|x| \leq \sqrt{2}} \varphi(x)\, dx ,$$

where α_t is the number of spheres, with centres at $x^{(k)}$ and radius $\sqrt{2}$, which are included in tM. Now

$$\int_{|x| \leq \sqrt{2}} \varphi(x)\, dx = \int_{|x| \leq \sqrt{2}} \left(1 - \frac{|x|^2}{2}\right) dx = n\sigma_n \int_0^{\sqrt{2}} \left(1 - \frac{r^2}{2}\right) r^{n-1} dr$$

$$= n\sigma_n \left(\frac{1}{n} \cdot 2^{n/2} - \frac{1}{n+2} \cdot 2^{n/2}\right) = \frac{\sigma_n \cdot 2^{n/2}}{\left(1 + \frac{n}{2}\right)} ,$$

so that

$$V_t \geq \frac{\alpha_t \cdot \sigma_n \cdot 2^{n/2}}{\left(1 + \frac{n}{2}\right)} , \quad \text{or} \quad \frac{\alpha_t \cdot \sigma_n}{V_t} \leq \left(1 + \frac{n}{2}\right) 2^{-n/2} .$$

As t tends to infinity, the difference

$$\frac{\alpha_t \cdot \sigma_n}{V_t} - \frac{\beta_t \cdot \sigma_n}{V_t} \to 0 ,$$

since $\alpha_t - \beta_t$ is the number of such of those centres as lie in a strip around the boundary, at a distance less than $\sqrt{2}$ from the boundary, and by the definition of Jordan volume, it is clear that

$$\frac{\text{volume of the strip}}{V_t} \to 0 \ .$$

We have therefore the desired result

$$\limsup_{t\to\infty} \frac{\beta_t \cdot \sigma_n}{V_t} \le \left(1 + \frac{n}{2}\right) 2^{-n/2} \ .$$

We still have to prove that the sum of the weights for any point is not greater than 1. Note that

$$
(20) \quad
\begin{aligned}
\left| x^{(k)} - x^{(j)} \right|^2 &= \left| x^{(k)} - x + x - x^{(j)} \right|^2 \\
&= \left| x^{(k)} - x \right|^2 + \left| x^{(j)} - x \right|^2 - 2\left(x^{(k)} - x \right) \cdot \left(x^{(j)} - x \right) \ .
\end{aligned}
$$

If we sum (20) over all values of j, k, with $j < k$, which are not greater than a given positive integer m, we get

$$
(21) \quad
\begin{aligned}
\sum_{1 \le j < k \le m} \left| x^{(k)} - x^{(j)} \right|^2 &= (m-1) \sum_{k=1}^{m} \left| x^{(k)} - x \right|^2 \\
&\quad - 2 \sum_{1 \le j < k \le m} \left(x^{(k)} - x \right) \cdot \left(x^{(j)} - x \right) \ .
\end{aligned}
$$

However

$$
(22) \quad \left| \sum_{k=1}^{m} \left(x^{(k)} - x \right) \right|^2 = \sum_{k=1}^{m} \left| x^{(k)} - x \right|^2 + 2 \sum_{1 \le j < k \le m} \left(x^{(k)} - x \right) \cdot \left(x^{(j)} - x \right) \ .
$$

Adding (21) and (22), we obtain

$$
(23) \quad \sum_{1 \le j < k \le m} \left| x^{(k)} - x^{(j)} \right|^2 + \left| \sum_{k=1}^{m} \left(x^{(k)} - x \right) \right|^2 = m \sum_{k=1}^{m} \left| x - x^{(k)} \right|^2 \ .
$$

Since the second term on the left-hand side is always non-negative, we have the following inequality:

$$
(24) \quad \sum_{1 \le j < k \le m} \left| x^{(k)} - x^{(j)} \right|^2 \le m \sum_{k=1}^{m} \left| x - x^{(k)} \right|^2 \ .
$$

Now suppose

$$\left| x^{(k)} - x^{(j)} \right| \ge 2 \ , \quad j \ne k \ ,$$

so that spheres of radius 1 do not overlap, and we find that the left-hand side of (24) is not less than $4m(m-1)/2$. Using this in (24), we have

$$
(25) \quad 2(m-1) \le \sum_{k=1}^{m} \left| x - x^{(k)} \right|^2 \ ,
$$

or

(26)
$$\sum_{k=1}^{m} \left(1 - \frac{\left| x - x^{(k)} \right|^2}{2} \right) \leq 1 .$$

To prove (16) we must show that

$$\psi(x) = \sum_{\substack{k \\ \left| x - x^{(k)} \right| \leq \sqrt{2}}} \left(1 - \frac{\left| x - x^{(k)} \right|^2}{2} \right) \leq 1 .$$

This follows from (26), since there can only be a finite number of terms in the sum for $\psi(x)$. We have therefore completed the proof of (15).

The bound we have obtained for q_n is not very good for small values of n. For example, for $n = 2$ the value is 1; for $n = 3$, the value is $\frac{5}{4\sqrt{2}}$, etc. We can obtain a refinement of the bound as follows.

Consider the points x, such that

$$\left| x - x^{(k)} \right| < 2 - \sqrt{2} .$$

x can be in no other sphere $|x - x^{(j)}| \leq \sqrt{2}$, $j \neq k$, for otherwise $|x^{(j)} - x^{(k)}| = |x - x^{(j)} + x^{(k)} - x| < 2$, which contradicts our assumption that the spheres of radius 1 do not overlap. Therefore, if

$$\left| x - x^{(k)} \right| < 2 - \sqrt{2} ,$$

then $|x - x^{(j)}| > \sqrt{2}$, $j \neq k$.

We now define a new weight-function $\tilde{\varphi}$, such that

$$\tilde{\varphi}(x) = \begin{cases} 1 & , & |x| < 2 - \sqrt{2} , \\ 1 - \frac{|x|^2}{2} & , & 2 - \sqrt{2} \leq |x| \leq \sqrt{2} , \\ 0 & , & |x| > \sqrt{2} . \end{cases}$$

The proof goes through with $\tilde{\varphi}$, and we find that

$$\limsup_{t \to \infty} \frac{\beta_t \cdot \sigma_n}{V_t} \leq \left(1 + \frac{n}{2} \right) 2^{-n/2} h_n ,$$

where

$$h_n^{-1} = 1 + \frac{n \left(\sqrt{2} - 1 \right)^{n+2}}{2} .$$

Unfortunately h_n approaches 1 as $n \to \infty$, so that the bounds are asymptotically equal.

Lecture XIII

§1. The Second Finiteness Theorem

Having completed our digression on the closest packing of spheres, we shall return to the theory of reduction. First we recall some of our notations.

Let $\mathcal{S}$ be the space of all symmetric $n \times n$ matrices, $\mathcal{P}$ the subspace of all positive-definite, symmetric $n \times n$ matrices and $\mathcal{R}$ that subspace of $\mathcal{P}$ which contains only reduced matrices. A positive-definite, symmetric $n \times n$ matrix $T = (t_{ij})$, whose associated quadratic form is $Q(x)$, is called reduced [see Lecture XI, § 2], if

$$(1) \qquad\qquad Q(g) \geq t_k \, ,$$

for all integral vectors g, such that

$$(2) \qquad \text{g.c.d.} \{g_k, g_{k+1}, \ldots, g_n\} = 1 \, , \quad (k = 1, \ldots, n),$$

and if, in addition,

$$(3) \qquad\qquad t_{1j} \geq 0 \, , \quad (j = 2, \ldots, n) \, .$$

In Lecture XI we saw that in case $n = 2$, or 3, the infinite set of conditions implied by (1) and (2) can be replaced by a finite set of conditions [see Lecture XI, §§ 4,6]. The Second Finiteness Theorem states that the same situation holds in the general case: *the infinite set of conditions given by (1) and (2) can be replaced by a finite set.*

This theorem is quite deep, and the proof of it will take this lecture and most of the next. A short review of the main features of the proof will help clarify the details.

In case $n = 2$, or 3, we saw that if we took g as one of a particular finite set of vectors $\{g^*\}$ – those whose coordinates are either -1, 0, or 1 – then the finite number of conditions

$$(4) \qquad\qquad Q(g^*) \geq t_k \, ,$$

where

$$(5) \qquad \text{g.c.d.} \{g_k^*, g_{k+1}^*, \ldots, g_n^*\} = 1 \, , \quad (k = 1, \ldots, n)$$

implies the infinite set of conditions given by (1) and (2). Our main aim in the general case will be to find a set of vectors similar to $\{g^*\}$.

Let $\mathcal{R}_U$ denote the space obtained by transforming all the matrices in $\mathcal{R}$ by the unimodular matrix U. That is to say, a matrix S belongs to $\mathcal{R}_U$, if there exists a reduced matrix T, such that

$$S = T[U] \ .$$

Since every matrix S in $\mathcal{P}$ is equivalent to a reduced matrix, it follows that S must belong to some $\mathcal{R}_U$. Therefore $\mathcal{P}$ is completely covered by the sets $\mathcal{R}_U$, where U runs over all unimodular matrices.

We introduce a family of subspaces of $\mathcal{P}$ which we denote by $\mathcal{P}_K$. If K is large enough – its value depends on n – we can show that $\mathcal{P}_K$ contains $\mathcal{R}$, and is covered by only a finite number of "images" $\mathcal{R}_U$. The column vectors of these matrices U will be the set of vectors $\{g^*\}$. Let $\mathcal{M} \subset \mathcal{S}$ be the space of matrices satisfying the finite number of conditions (4) and (5) and also (3). We complete our proof of the Second Finiteness Theorem by showing that the space $\mathcal{M} \cap \mathcal{P}$ is identical with the space $\mathcal{R}$.

§2. An inequality for positive-definite symmetric matrices

We shall need a great deal of preliminary work, before we can define the space $\mathcal{P}_K$. First we obtain an important inequality.

Let S be a positive-definite, symmetric $n \times n$ matrix with elements s_{ij} and determinant Δ. Denote the elements along the principal diagonal of S by $s_1, \ldots, s_n$. Then we shall prove that [see Lecture XI, (5)]

(6) $s_1 \ldots s_n \geq \Delta \ .$

The equality sign holds only if S is a diagonal matrix.

The proof is by induction. For $n = 1$ the inequality is trivial. For $n > 1$, let Δ_1 represent the minor of S corresponding to s_1. We shall prove that

(7) $s_1 \Delta_1 \geq \Delta \ ,$

and that the equality holds only if all the elements of the first row and column, except s_1, are zero. Inequality (7) contains inequality (6), since by induction if we assume (6) for $n - 1$, that is

$$s_2 \ldots s_n \geq \Delta_1 \ ,$$

and combine this with (7), we get

$$s_1 s_2 \ldots s_n \geq s_1 \Delta_1 \geq \Delta \ .$$

We prove (7) by expanding Δ in terms of elements of the first row and column. We have

(8)
$$\Delta = s_1\Delta_1 - \sum_{i,j=2}^{n} S_{ij} \cdot s_{1i} \cdot s_{j1} ,$$

where S_{ij} is the cofactor in $S_{[1]}$ corresponding to s_{ij}, where $S_{[1]}$ is the matrix obtained from S by erasing the first row and the first column. The matrix whose elements are S_{ij} is the adjoint of $S_{[1]}$. The positive-definiteness of S implies that of $S_{[1]}$, which in turn implies that of its adjoint. It follows that the quadratic form whose matrix has coefficients S_{ij} is positive-definite. Therefore

$$\Delta \leq s_1\Delta_1 ,$$

and we have equality only if

$$s_{1i} = s_{j1} = 0 , \quad (i \neq 1, j \neq 1) .$$

This proves inequality (7).

§3. The space $\mathcal{P}_K$

We wish to define a space $\mathcal{P}_K$, which will contain the space $\mathcal{R}$. We have proved in Lecture XI that if a matrix T belongs to $\mathcal{R}$, then the elements of T satisfy the following inequalities:

(9) $$t_1 \ldots t_n \leq k \cdot \det T \quad , \qquad \text{[XI,(4)]}$$

(10) $$t_1 \leq t_2 \leq \ldots \leq t_n \quad , \qquad \text{[XI,(9)]}$$

(11) $$|2t_{ij}| \leq t_i , \quad (i < j) , \qquad \text{[XI,(9)]}$$

where

$$k = \left(\frac{4}{\pi}\right)^n \left(\frac{3}{2}\right)^{n(n-1)} \left[\Gamma\left(\frac{n}{2}+1\right)\right]^2 .$$

We shall say that a matrix $S \in \mathcal{P}$ belongs to $\mathcal{P}_K$, if its elements satisfy the following inequalities, which are slightly more general than (9), (10), or (11):

(A) $$s_1 \ldots s_n < K \cdot \det S ,$$

(B) $$s_i < K s_{i+1} , \quad (i = 1, \ldots, n - 1) ,$$

(C) $$|s_{ij}| < K s_i , \quad (i < j) .$$

Note that from (A) we have $\mathcal{P}_K = \emptyset$ for $K \leq 1$. *We shall therefore assume that* $K > 1$.

Our aim in this lecture will be to prove that if S and R are matrices belonging to $\mathcal{P}_K$, and if $R = S[U]$, where U is unimodular, then U is bounded, that is, *all* its elements are bounded. The proof of this theorem will depend on four lemmas.

Lemma 1. *Let* $S \in \mathcal{P}$, *and* $S[x]$, $x = \begin{pmatrix} x_1 \\ \vdots \\ x_n \end{pmatrix}$, *be the associated quadratic form.*
If S *satisfies condition* (A), *then*

$$(12) \qquad K^{-1}C^{1-n} < \frac{S[x]}{s_1 x_1^2 + \ldots + s_n x_n^2} < C ,$$

for all $x \neq 0$, *where* C *depends only on* n, *and is independent of* K.

If we replace $x_1, \ldots, x_n$ by $s_1^{-1/2} x_1, \ldots, s_n^{-1/2} x_n$, then (12) may be written
as

$$(13) \qquad K^{-1}C^{1-n} < \frac{S^*[x]}{x_1^2 + \ldots + x_n^2} < C ,$$

where S^* is the matrix whose elements are $s_{ij}^* = s_i^{-1/2} s_{ij} s_j^{-1/2}$. We have

$$(14) \qquad \det S^* = \frac{\det S}{s_1 \ldots s_n} .$$

Note also that the diagonal elements of S^* are all equal to 1. Condition (A)
then becomes

$$(15) \qquad 1 < K \cdot \det S^* .$$

Since the ratio

$$\frac{S^*[x]}{x_1^2 + \ldots + x_n^2}$$

is homogeneous of degree zero, we may always assume that

$$(16) \qquad x_1^2 + \ldots + x_n^2 = 1 .$$

We now investigate the minimum and maximum values attained by $S^*[x]$
on the unit sphere (16). Denote the minimum by α and the maximum by β.
We must prove that

$$(17) \qquad \beta < C ,$$

and that

$$(18) \qquad \alpha > K^{-1}C^{1-n} ,$$

where C depends only on n.
It is well known that α and β are the smallest and largest roots, respectively,
of the characteristic equation

$$(19) \qquad \det(\lambda E - S^*) = 0 ,$$

where E is the $n \times n$ unit matrix.

Since S^* is positive-definite, all the roots of equation (19) are positive. Equation (19), when expanded in powers of λ, is

(20)·
$$\lambda^n + a_1\lambda^{n-1} + \ldots + a_n = 0 ,$$

where the coefficients are algebraic sums of minors of S^*.

Since the diagonal elements of S^* are all 1, we can show that all the other elements will be less, in absolute value, than 1. For consider the quadratic form in which $x_k = 0$ for $k \neq i$, $k \neq j$, $i \neq j$; then

(21)
$$0 < S^*[x] = x_i^2 + x_j^2 + 2s_{ij}^* x_i x_j , \quad \text{for } (x_i, x_j) \neq (0,0) .$$

Hence
$$1 - \left(s_{ij}^*\right)^2 > 0 , \quad \text{or} \quad \left(s_{ij}^*\right)^2 < 1 .$$

Now, since all the elements of S^* are bounded, all the minors formed from it will be bounded, and therefore all the coefficients in (20) are bounded. This implies that all the roots of (20) are bounded by some constant C, depending only on n.

The product of all the roots is $\det S^*$, where

(22)
$$\det S^* > K^{-1} ,$$

by (15). Since all the roots are less than C, we have

(23)
$$\det S^* < C^{n-1}\lambda_k ,$$

for each root λ_k. In particular, $\det S^* < C^{n-1}\alpha$, and combining this with (22), we have

(24)
$$\alpha > K^{-1}C^{1-n} .$$

This completes the proof of Lemma 1.

Lemma 2. *Let $S, R \in \mathcal{P}$. Suppose that S and R both satisfy conditions (A) and (B), and that they are equivalent; that is, there exists a unimodular matrix U such that $S[U] = R$. Then we have*

(25)
$$s_i < r_i K^n C^{n-1} , \quad \text{for } i = 1, \ldots, n ,$$

and, by symmetry, we also have

(26)
$$r_i < s_i K^n C^{n-1} , \quad \text{for } i = 1, \ldots, n .$$

Let the elements of the jth column $u^{(j)}$ of U be denoted by u_{ij} (that is, $u_i^{(j)} = u_{ij}$), then

(27)
$$r_j = S[u^{(j)}] .$$

Here $u^{(j)}$ cannot be the zero vector, for otherwise U would not be unimodular. We can now apply Lemma 1, and obtain

$$K^{-1}C^{1-n} < \frac{r_j}{(s_1 u_{1j}^2 + \ldots + s_n u_{nj}^2)} \, ,$$

or

(28) $$s_1 u_{1j}^2 + \ldots + s_n u_{nj}^2 < r_j K C^{n-1} \, .$$

By condition (B),

$$r_j \leq r_k K^{k-j} \, , \quad (j \leq k) \, ,$$

so that on using (28), we have

(29) $$s_1 u_{1j}^2 + \ldots + s_n u_{nj}^2 < r_k K^{k-j+1} C^{n-1} \, , \quad (j \leq k) \, .$$

If we could show that for all k there exists a non-zero element u_{hj}, where

(30) $$h \geq k \geq j \, ,$$

we would be finished, because (29) implies

(31) $$s_h u_{hj}^2 < r_k K^{k-j+1} C^{n-1} \, ,$$

or

(32) $$s_h < r_k K^{k-j+1} C^{n-1} \, ,$$

since $u_{hj} \neq 0$ is an element of a unimodular matrix, therefore an integer, and therefore not less than 1 in absolute value. Since S satisfies condition (B), we have

$$s_h \geq s_k K^{k-h} \, , \quad \text{since } k \leq h \, ,$$

and on using (32), we get the desired result:

(33) $$s_k < r_k K^{h-j+1} C^{n-1} \leq r_k K^n C^{n-1} \, ,$$

since $h \leq n$, $j \geq 1$, and $K > 1$. It remains only to prove that there exists a non-zero element u_{hj}, where $h \geq k \geq j$; that is, a non-zero element in the shaded region of $n - k + 1$ rows and k columns in the matrix U.

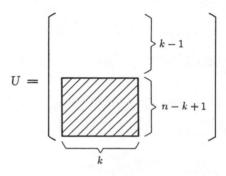

If we expand the determinant of U, by Laplace's theorem for the first $k - 1$ rows, we get an algebraic sum of products of $(k - 1)^{th}$ order minors by $(n - k + 1)^{th}$ order minors. Now if all the elements in the shaded portion of U were zero, then every $(n - k + 1)^{th}$ order submatrix (corresponding to a minor of the same order in the Laplace expansion) would contain a zero column, and its determinant would be zero. This is a consequence of the fact that $(n - k + 1) + k = n + 1$. If all these minors were zero, the Laplace expansion would be zero, and the determinant of U would be zero. This contradicts our assumption that U is unimodular; therefore, there exists a non-zero element u_{hj}, where $h \geq k \geq j$. And this completes the proof of Lemma 2.

Lemma 3. *Under the same assumptions as in Lemma 2, we shall prove that there exists a q, $1 \leq q \leq n$, such that U has the following form*

$$(34) \qquad U = \begin{pmatrix} U_1 & U_{12} \\ 0 & U_2 \end{pmatrix} ,$$

where U_1 is a $(q - 1) \times (q - 1)$ matrix, and U_2 is a $(n - q + 1) \times (n - q + 1)$ matrix. Moreover, U_2 is bounded.

The value of q will depend upon the matrices S and R. If $q = 1$, the lemma means that the whole matrix is bounded. If $q = n$, it means that u_{nn} is bounded, while all other elements in the last row are zero.

We define q as follows: for all $k \geq q$, we have

$$(35) \qquad s_{k+1} \leq r_k (KC)^{n-1} ,$$

while

$$(36) \qquad s_q > r_{q-1}(KC)^{n-1} ,$$

where we set $s_{n+1} = 0$, and $r_0 = 0$.

We had from (28)

$$(37) \qquad s_1 u_{1j}^2 + \ldots + s_n u_{nj}^2 < r_j KC^{n-1} ,$$

where $j = 1, \ldots, n$. This gives, for each h, $1 \leq h \leq n$,

$$s_h u_{hj}^2 < r_j KC^{n-1} ,$$

or

$$(38) \qquad u_{hj}^2 < r_j KC^{n-1} s_h^{-1} .$$

From (35) and Lemma 2 it follows that s_{k+1}/s_k is bounded for $k \geq q$; but then

$$r_q, \ldots, r_n, s_q, \ldots, s_n$$

all have the same order of magnitude. That is, $r_j s_h^{-1}$ is bounded if j, h lie between q and n. From (38) it follows that u_{hj}^2 is bounded if we have the condition

$$q \leq h \leq n , \quad q \leq j \leq n ;$$

that is to say, U_2 is bounded.

Suppose now that $h \geq q$ and $j \leq q - 1$. The inequalities

$$r_j \leq r_{q-1} K^{q-j-1} ,$$

and

$$s_h \geq s_q K^{q-h} ,$$

follow from condition (B). We therefore get from (38)

$$u_{hj}^2 < r_{q-1} K^{q-j-1} \cdot K C^{n-1} \cdot K^{h-q} s_q^{-1}$$
$$= r_{q-1} K^{h-j} C^{n-1} s_q^{-1} \leq r_{q-1}(KC)^{n-1} s_q^{-1} ,$$

but, by (36),

$$r_{q-1} s_q^{-1} < (KC)^{1-n} ,$$

therefore $u_{hj}^2 < 1$. This implies that $u_{hj} = 0$, since u_{hj} is an integer.

Lemma 4. *Let $S, R \in \mathcal{P}$. Suppose that $S[U] = R$, with U unimodular, and that R and S satisfy conditions (A),(B),(C), so that they both belong to $\mathcal{P}_K$. Then U is bounded, and the bound depends only on n and K.*

This is our main lemma. It shows that there are only a finite number of unimodular matrices which transform a matrix of $\mathcal{P}_K$ into another matrix of $\mathcal{P}_K$.

The proof depends on Lemma 3. Since R and S both satisfy conditions (A) and (B), Lemma 3 states that

$$(39) \qquad\qquad U = \begin{pmatrix} U_1 & U_{12} \\ 0 & U_2 \end{pmatrix} ,$$

where U_1 has certainly less than n rows. (Actually U_1 is a $(q-1) \times (q-1)$ matrix for some q, $1 \leq q \leq n$.) We split S and R similarly to the way U is split. We write

$$(40) \qquad\qquad S = \begin{pmatrix} S_1 & S_{12} \\ \cdot & \cdot \end{pmatrix} , \qquad R = \begin{pmatrix} R_1 & R_{12} \\ \cdot & \cdot \end{pmatrix} .$$

Note that

$$U' = \begin{pmatrix} U_1' & 0 \\ U_{12}' & U_2' \end{pmatrix}$$

so that

$$(41) \qquad U'SU = \begin{pmatrix} U_1' & 0 \\ U_{12}' & U_2' \end{pmatrix} \begin{pmatrix} S_1 & S_{12} \\ \cdot & \cdot \end{pmatrix} \begin{pmatrix} U_1 & U_{12} \\ 0 & U_2 \end{pmatrix} = \begin{pmatrix} R_1 & R_{12} \\ \cdot & \cdot \end{pmatrix} .$$

Performing the multiplication in (41), we find that

$$(42) \qquad R_1 = U_1' S_1 U_1 , \qquad R_{12} = U_1' S_1 U_{12} + U_1' S_{12} U_2 .$$

We may write

$$U_1' S_1 U_{12} = \underbrace{U_1' S_1 U_1} \, U_1^{-1} U_{12} = R_1 U_1^{-1} U_{12} ,$$

$$U_1' S_{12} U_2 = U_1' S_1 U_1 U_1^{-1} S_1^{-1} S_{12} U_2 = R_1 U_1^{-1} S_1^{-1} S_{12} U_2 ,$$

so that

$$R_{12} = R_1 U_1^{-1} U_{12} + R_1 U_1^{-1} S_1^{-1} S_{12} U_2 .$$

Solving for U_{12}, we find that

$$(43) \qquad U_{12} = U_1 R_1^{-1} R_{12} - S_1^{-1} S_{12} U_2 .$$

We shall prove, by induction, that U is bounded. For $n = 1$, U_1 is bounded, since $n - q + 1 = 1$. Suppose that we have proved that U_1 is bounded for the cases $1, \ldots, n - 1$. By (42) we have

$$R_1 = S_1[U_1] .$$

Now if R_1 and S_1 belonged to the set $\mathcal{P}_K$ for $(q - 1) \times (q - 1)$ matrices, we could conclude that U_1 is bounded. We must prove only that

$$\varrho = s_1 \ldots s_{q-1} (\det S_1)^{-1} < K ,$$

since conditions (B) and (C) are satisfied by hypothesis. Multiplying and dividing ϱ by the product $s_q \ldots s_n$, we have

$$(44) \qquad \varrho = s_1 \ldots s_n (\det S_1)^{-1} (s_q \ldots s_n)^{-1} .$$

But by using (7) repeatedly, we get

$$(45) \qquad \det S \leq s_n \Delta_1 \leq s_n s_{n-1} \Delta_{12} \leq \ldots \leq s_n \ldots s_q \det S_1 ,$$

where Δ_1 is the minor of S corresponding to s_n, Δ_{12} the minor corresponding to s_n, s_{n-1}, etc. Using (45) in (44), we get

$$\varrho \leq s_1 \ldots s_n (\det S)^{-1} < K ,$$

since S satisfies condition (A). This proves that R_1 and S_1 belong to $\mathcal{P}_K$ for $(q - 1) \times (q - 1)$ matrices, and therefore U_1 is bounded.

Now if we can prove that $R_1^{-1} R_{12}$ and $S_1^{-1} S_{12}$ are bounded, then (43) would imply that U_{12} is bounded, and we would be finished with the proof of Lemma 4.

Let σ_{ij} be the cofactor of s_{ij} in the matrix S_1, and ϱ_{ij} represent the elements of the matrix S_1^{-1}. It is clear that

$$(46) \qquad \varrho_{ij} = \sigma_{ij} (\det S_1)^{-1} .$$

Now σ_{ij} is an algebraic sum of products of terms from each column of S_1 except the jth. Since, by condition (C), and the fact that $K > 1$,

$$|s_{ij}| < Ks_i, \quad (i \le j),$$

and by symmetry,

$$|s_{ij}| = |s_{ji}| < Ks_j, \quad (j \le i),$$

and by condition (B) every element of S_1 has at most the order of magnitude of the diagonal element in the same column, we find that

$$|\sigma_{ij}| < C_2(n, K) \cdot s_1 \ldots s_{q-1} s_j^{-1},$$

where $C_2(n, K)$ is a constant depending only on n and K.

Using (46), we have

$$(47) \qquad |\varrho_{ij}| \le C_2(n, K) \cdot s_1 \ldots s_{q-1} s_j^{-1} (\det S_1)^{-1} < C_3(n, K) \cdot s_j^{-1}.$$

If we denote the elements of S_{12} by τ_{ij}, the absolute value of a general element of $S_1^{-1} S_{12}$ will be

$$(48) \qquad \left| \sum_{k=1}^{q-1} \varrho_{ik} \tau_{kj} \right| < \sum_{k=1}^{q-1} C_3 s_k^{-1} C_4 s_k < C_5,$$

where we have used (47) and condition (C). This shows that all the elements of $S_1^{-1} S_{12}$ are bounded. Since R satisfies the same conditions as S does, we conclude that $R_1^{-1} R_{12}$ is also bounded. Therefore, from (43), U_{12} is bounded, and we have proved Lemma 4.

§4. Images of $\mathcal{R}$

We showed in Lectures X and XI that given any positive-definite, symmetric matrix S, there exists a unimodular matrix U, such that $T = S[U]$ is reduced, that is, T belongs to $\mathcal{R}$. We can express S in terms of T as follows:

$$S = T[U^{-1}].$$

This relation implies that if we apply all possible unimodular transformations to $\mathcal{R}$, we shall get the space $\mathcal{P}$, or stated differently, any point in $\mathcal{P}$ is the image of a point in $\mathcal{R}$.

Let $\mathcal{R}_U$ represent the image of $\mathcal{R}$ under the unimodular transformation U. We have seen that the set of images $\{\mathcal{R}_U\}$ covers $\mathcal{P}$. We wish to show that the images $\mathcal{R}_U$ do not overlap, and finally that $\mathcal{R}$ is bounded by a finite number of planes.

We first prove that $\mathcal{P}_K$ contains $\mathcal{R}$ if $K > (\frac{4}{\pi})^n (\frac{3}{2})^{n(n-1)} [\Gamma(\frac{n}{2} + 1)]^2$. To do this we must show that a reduced matrix satisfies conditions (A), (B), (C).

The First Finiteness Theorem, applied to reduced quadratic forms [Lecture XI, § 2] states that

$$t_1 \ldots t_n \leq \left(\frac{4}{\pi}\right)^n \left(\frac{3}{2}\right)^{n(n-1)} \left[\Gamma\left(\frac{n}{2}+1\right)\right]^2 \det T \; ;$$

that is to say, condition (A) is satisfied. We also proved [in Lecture XI, § 3] that for reduced forms

$$t_i \leq t_{i+1} , \quad i = 1, \ldots, n-1 ;$$

$$|2t_{ij}| \leq t_i , \quad i < j ,$$

so that conditions (B) and (C) are satisfied.

Since the set of images $\{\mathcal{R}_U\}$, with U unimodular covers all of $\mathcal{P}$, it is trivial that it also covers $\mathcal{P}_K$. We prove more: *the space $\mathcal{P}_K$ is covered by a finite number of images $\mathcal{R}_U$.*

Suppose S is in $\mathcal{P}_K$. Then S is covered by some $\mathcal{R}_U$; that is to say

$$S = T[U] ,$$

where T is reduced. Since $\mathcal{R}$ is a part of $\mathcal{P}_K$, it follows that T belongs to $\mathcal{P}_K$, and then by Lemma 4, U is bounded. Since the elements of U are integers, and are bounded, there are only a finite number of possibilities for the elements of U, and so the number of images must be finite.

Lecture XIV

§1. Boundary points

A *boundary point* of $\mathcal{R}$ is a point in $\mathcal{S}$, such that arbitrarily near to it (in the sense of the Euclidean distance in the space $\mathcal{S}$) there exist points belonging to $\mathcal{R}$ and points not belonging to $\mathcal{R}$. [Notation as in § 1 of Lecture XIII.] A boundary point of $\mathcal{R}$ may not belong to $\mathcal{P}$; for example, the zero matrix does not belong to $\mathcal{P}$, yet it is a boundary point of $\mathcal{R}$, because λT (λ arbitrary positive) belongs to $\mathcal{R}$ if T does, and we may let λ tend to zero.

If, however, the boundary point belongs to $\mathcal{P}$, it must belong to $\mathcal{R}$. This is trivial since, as the points of $\mathcal{R}$ approach the boundary point, the reduction conditions may go over into equalities, but this is still permitted.

Recalling the reduction conditions [Lecture XIII, § 1], we note that $S \in \mathcal{R}$ if and only if $S \in \mathcal{P}$, and

$$(*) \qquad\qquad S[g] \geq s_i \, ,$$

for all integral vectors $g = \begin{pmatrix} g_1 \\ \vdots \\ g_n \end{pmatrix}$, such that g.c.d. $\{g_i, g_{i+1}, \ldots, g_n\} = 1$, $(i = 1, \ldots, n)$, and

$$(**) \qquad\qquad s_{1i} \geq 0 \, , \quad (i = 2, \ldots, n) \, .$$

Note that condition $(*)$ may be trivial. That is the case if and only if g is plus or minus the ith unit vector, so that condition $(*)$ becomes $s_i \geq s_i$. In the sequel we mean by a *reduction condition* one which is different from these trivial ones.

If for the point S in $\mathcal{R}$, at least one of the reduction conditions becomes an equality, then S is a boundary point of $\mathcal{R}$. The proof is simple. The reduction condition which S makes an equality is a homogeneous linear inequality; therefore arbitrarily near to S there exist points T, such that the condition becomes an incquality in the wrong direction. This means that T does not lie in $\mathcal{R}$ and so S must be a boundary point of $\mathcal{R}$.

§ 2. Non-overlapping of images

We shall now prove that different images of $\mathcal{R}$ do not have a common neighbourhood. Note first that

$$T[U] = T[-U] \,,$$

so that $\mathcal{R}_U = \mathcal{R}_{-U}$. We must therefore *identify U and $-U$*.

Suppose now that $\mathcal{R}_U$ and $\mathcal{R}_V$ do overlap, where $V \neq U$, $V \neq -U$, and U, V are unimodular. If we apply the mapping V^{-1} to both regions, we still have an overlapping. Therefore $\mathcal{R}_W$ and $\mathcal{R}$ overlap, where $W = UV^{-1} \neq \pm E$, E being the identity matrix.

Let S be a common point of $\mathcal{R}$ and $\mathcal{R}_{W^{-1}}$; that is, S belongs to $\mathcal{R}$, and also $T = S[W]$, where T belongs to $\mathcal{R}$. Let $w^{(k)}$ represent the kth column vector of W. There are two cases to be considered.

First case. It may happen that

$$w^{(k)} = \pm e^{(k)} \,,$$

for all k, where $e^{(k)}$ is the kth unit vector (i.e. the kth column vector of E).

Since W is not $\pm E$, we must have at least two columns of opposite signs. Replacing W by $-W$ if necessary, we may assume that

(1) $$w^{(1)} = e^{(1)}, \; \ldots \;, \quad w^{(i-1)} = e^{(i-1)} \,, \quad w^{(i)} = -e^{(i)} \,.$$

We shall prove that $s_{1i} = 0$, so that S is a boundary point of $\mathcal{R}$. We have [by $(**)$ of § 1]

$$t_{1i} \geq 0 \,, \quad \text{and } s_{1i} \geq 0 \,, \quad \text{since } i > 1 \,.$$

Moreover, $t_{1i} = \left(e^{(1)}\right)' Te^{(i)} = \left(e^{(1)}\right)' W'SWe^{(i)} = \left(e^{(1)}\right)' S\left(-e^{(i)}\right) = -s_{1i}$. Hence $s_{1i} = 0$.

Second case. Suppose not all columns of W are unit vectors. More precisely assume that

(2) $$w^{(1)} = \pm e^{(1)}, \; \ldots \;, \quad w^{(i-1)} = \pm e^{(i-1)} \,, \quad w^{(i)} \neq \pm e^{(i)} \,.$$

We now apply condition $(*)$ of § 1:

$$S[w^{(i)}] \geq s_i \,.$$

It is clear that this condition must hold for $w^{(i)}$, because otherwise W would not be unimodular. Now $t_i = S[w^{(i)}]$, so we have $t_i \geq s_i$. But since $S = T[W^{-1}]$, we also have $s_i \geq t_i$, since the first $i - 1$ columns of W^{-1} are also the first $i - 1$ unit vectors, except for sign, the greatest common divisor of the last $n - i + 1$ components of the ith column vector of W^{-1} is equal to 1. Therefore $s_i = t_i$, and the reduction condition reduces to the equality

(3) $$S[w^{(i)}] = s_i \,,$$

which is not trivial, since $w^{(i)} \neq \pm e^{(i)}$. Again this implies that S is a boundary point of $\mathcal{R}$.

Consider all unimodular matrices W which are not diagonal and which are such that $\mathcal{R} \cap \mathcal{R}_W \neq \emptyset$ (or equivalently $\mathcal{R} \cap \mathcal{R}_{W^{-1}} \neq \emptyset$). Since W and $-W$ have to be identified, choose one representative of these two. We thus obtain a set $\mathcal{W}$ of unimodular matrices. Note that $\mathcal{W}$ is a finite set, since, by definition, for $W \in \mathcal{W}$ there exist $S, T \in \mathcal{R}$, such that $T = S[W]$; and $\mathcal{R}$ is contained in $\mathcal{P}_K$ for K sufficiently large, so that, by Lemma 4 of Lecture XIII, W is bounded independently of S and T.

For each $W \in \mathcal{W}$, take the corresponding column vector $w^{(i)}$, as was done in the 'second case' above. Let $\mathcal{W}^{(i)}$ denote the set of all such vectors as correspond to the same integer i, $i = 1, \ldots, n$.

§3. Space defined by a finite number of conditions

We have proved that any point common to $\mathcal{R}$ and $\mathcal{R}_U$, where $U \neq \pm E$, U unimodular, makes at least one of the reduction conditions an equality, and must therefore be a boundary point of $\mathcal{R}$. We can also say that this boundary point S must lie on some of the finite number of planes given by the following equations:

(4) $$s_{1j} = 0 , \quad j = 2, \ldots, n ,$$

(5) $$S[w^{(i)}] = s_i , \quad w^{(i)} \in \mathcal{W}^{(i)} , \quad i = 1, \ldots, n .$$

Note that (5) is the equation of a plane, since it is a non-trivial linear relation connecting the elements of S. The number of planes is finite, since we showed that there were only a finite number of possibilities for $w^{(i)}$.

An illustration for the case $n = 2$ may clarify the situation. S is the three-dimensional space of all matrices of the form

$$\begin{pmatrix} a & b \\ b & c \end{pmatrix} .$$

$\mathcal{P}$ is the space of all such matrices for which $a > 0$ and $ac - b^2 > 0$.

As the set $\mathcal{W}$ we may take

$$\mathcal{W} = \left\{ \begin{pmatrix} 1 & 1 \\ 0 & -1 \end{pmatrix} , \begin{pmatrix} 0 & -1 \\ 1 & 1 \end{pmatrix} , \begin{pmatrix} 1 & 0 \\ -1 & -1 \end{pmatrix} , \right.$$
$$\left. \begin{pmatrix} 1 & 1 \\ -1 & 0 \end{pmatrix} , \begin{pmatrix} 0 & 1 \\ 1 & 0 \end{pmatrix} , \begin{pmatrix} 0 & -1 \\ 1 & 0 \end{pmatrix} \right\} .$$

This gives rise to

$$\mathcal{W}^{(1)} = \left\{ \begin{pmatrix} 0 \\ 1 \end{pmatrix} , \begin{pmatrix} 1 \\ -1 \end{pmatrix} \right\} , \quad \text{and} \quad \mathcal{W}^{(2)} = \left\{ \begin{pmatrix} 1 \\ -1 \end{pmatrix} \right\} .$$

The finite number of planes on which a boundary point of $\mathcal{R}$, which also lies in some $\mathcal{R}_U$ (for U unimodular, $U \neq \pm E$), must lie are given by the equations

(6)
$$b = 0 \, ,$$
$$c = a \, ,$$
$$a - 2b + c = a \, , \quad \text{or} \quad c = 2b \, ,$$
$$a - 2b + c = c \, , \quad \text{or} \quad a = 2b \, .$$

We have proved before [see Lecture XI, § 4] that $\mathcal{R}$ is the space defined by

(7)
$$b \geq 0 \, , \quad a \geq 2b \, , \quad c \geq a \, , \quad a > 0 \, .$$

Let us return to the general case. Suppose that we have a point S in $\mathcal{P}$ which is a boundary point of $\mathcal{R}$. Then S belongs to $\mathcal{R}$. We want to show that S is also in some $\mathcal{R}_W$, for $W \neq \pm E$, W unimodular. By the previous argument, S then satisfies at least one of the equations given in (4) and (5). We have seen that for K sufficiently large, we have $\mathcal{P}_K \supset \mathcal{R}$. Since $\mathcal{P}_K$ was defined by strict inequalities as a subset of $\mathcal{P}$, which is open in $\mathcal{S}$ [see Lecture XI, § 1], $\mathcal{P}_K$ is an open set in $\mathcal{S}$. Since S is a boundary point of $\mathcal{R}$, we may choose a sequence of points in $\mathcal{P}_K$, but not in $\mathcal{R}$, which converges to S. We have seen [in Lecture XIII] that $\mathcal{P}_K$ is covered by a finite number of $\mathcal{R}_U$'s. Every point of the chosen sequence lies therefore in the *finite union* $\mathcal{U}$, say, of those $\mathcal{R}_U$'s, where $U \neq \pm E$. Since every $\mathcal{R}_U$ is closed in $\mathcal{P}$, $\mathcal{U}$ is closed in $\mathcal{P}$, and since the sequence converges to $S \in \mathcal{P}$, we have $S \in \mathcal{U}$. It follows that there is a unimodular $W \neq \pm E$, such that $S \in \mathcal{R}_W$.

Consider the space $\mathcal{M}$ defined by the following inequalities, corresponding to the equalities (4) and (5): $S \in \mathcal{M}$ if and only if $S \in \mathcal{S}$, and

(8)
$$s_{1j} \geq 0 \, , \quad j = 2, \ldots, n \, ,$$

(9)
$$S[w^{(i)}] \geq s_i \, , \quad w^{(i)} \in W^{(i)} \, , \quad i = 1, \ldots, n \, .$$

We shall prove that *every point of $\mathcal{M}$ belongs to $\mathcal{R}$ or to its boundary*. Since $\mathcal{M}$ is defined by a finite number of inequalities, the *Second Finiteness Theorem* will follow.

§ 4. The Second Finiteness Theorem

The proof will be in two parts. First we show that if T is a point in $\mathcal{P}$ satisfying (8) and (9), then T belongs to $\mathcal{R}$. Secondly, if T satisfies (8) and (9) but is not in $\mathcal{P}$, then T is a boundary point of $\mathcal{R}$.

The space $\mathcal{P}_K$ is an open set in $\mathcal{S}$. It contains, for example, for $K > 1$, the unit matrix, and must also contain a sphere about it, so that the unit matrix is an interior point. The space $\mathcal{R}$ must also contain interior points, for if it had no interior points, the union of a finite number of images $\mathcal{R}_U$ would not contain interior points; but a finite number of images $\mathcal{R}_U$ cover $\mathcal{P}_K$. This contradiction shows that $\mathcal{R}$ contains interior points, and so it must contain a point R not lying on any of the planes given by the equality sign in (8), (9).

Now let T be a point in $\mathcal{P}$ satisfying (8) and (9). We will prove that T belongs to $\mathcal{R}$. Consider the straight line segment

$$\{T_\lambda = (1 - \lambda)R + \lambda T \,|\, 0 \leq \lambda \leq 1\} \ .$$

Since we proved that $\mathcal{P}$ is a convex cone [see Lecture XI, § 1], it follows that T_λ belongs to $\mathcal{P}$ for $0 \leq \lambda \leq 1$. Note that $T_0 = R$, $T_1 = T$.

If T does not belong to $\mathcal{R}$, it will be an exterior point $[T \in \mathrm{Int}(\mathcal{P} - \mathcal{R})]$ of $\mathcal{R}$, since $\mathcal{R}$ is closed with respect to $\mathcal{P}$. Therefore there must exist a value of λ, where $0 < \lambda < 1$, such that T_λ is a boundary point of $\mathcal{R}$.

We now show that this leads to a contradiction. Since R is an interior point of $\mathcal{R}$, it satisfies (8) and (9) with the strict inequality, and since T also satisfies (8) and (9), it follows that

$$T_\lambda = (1 - \lambda)R + \lambda T$$

will satisfy all of (8) and (9) in the strict sense if $0 \leq \lambda < 1$; therefore T_λ cannot be a boundary point unless $\lambda = 1$.

We have proved that any point common to $\mathcal{M}$ and $\mathcal{P}$ is a point in $\mathcal{R}$. We shall prove that if a point T is in $\mathcal{M}$ but not in $\mathcal{P}$, then T is a boundary point of $\mathcal{R}$.

Again let R be an interior point of $\mathcal{R}$, and consider

$$T_\lambda = (1 - \lambda)R + \lambda T \ , \quad 0 \leq \lambda \leq 1 \ .$$

Suppose that the elements of T_λ are represented by t_{ij}^λ.

As λ goes from 0 to 1, T_λ goes from R, an interior point of $\mathcal{R}$, to T which is not in $\mathcal{R}$. There exists a smallest value of λ, say λ_0, for which T_{λ_0} is a boundary point of $\mathcal{R}$. We shall show that $T_{\lambda_0} = T$, so that T is a boundary point of $\mathcal{R}$.

Since T_λ is an interior point of $\mathcal{R}$ for $\lambda < \lambda_0$, we have

(10) $$t_1^\lambda \leq t_2^\lambda \leq \ldots \leq t_n^\lambda \ ,$$

$$t_1^\lambda \ldots t_n^\lambda < K \cdot \det T_\lambda \ ,$$

(11) $$0 \leq t_{1i}^\lambda \leq \frac{t_1^\lambda}{2} \ , \quad (i > 1) \ .$$

We prove that $t_1^\lambda \to 0$ as $\lambda \uparrow \lambda_0$.

We proved before [see Lecture XIII, Lemma 1] that if $S \in \mathcal{P}$ satisfies $s_1 \ldots s_n < K \cdot \det S$, we have

(12) $$s_1 x_1^2 + \ldots + s_n x_n^2 \leq K' S[x] \ ,$$

for some constant K', depending only on K and n. This inequality will be satisfied by T_λ for $\lambda < \lambda_0$ and also in the limit by T_{λ_0}. But for T_{λ_0} we have

$$K'T_{\lambda_0}[x] \geq t_1^{\lambda_0} x_1^2 + \ldots + t_n^{\lambda_0} x_n^2 \, ,$$

and if $t_1^{\lambda_0}, \ldots, t_n^{\lambda_0}$ were all positive, T_{λ_0} would belong to $\mathcal{P}$. This leads to a contradiction by the same argument as in the first part of the proof.

Because of (10) and since $t_1^{\lambda_0}$ is the limit of t_1^λ for $\lambda \uparrow \lambda_0$, we must have $t_1^{\lambda_0} = 0$, and by (11) $t_{1i}^{\lambda_0} = 0$, $i > 1$. Since R is an interior point of $\mathcal{R}$, we have $r_{1i} > 0$, $i > 1$, and since T satisfies (8) and (9) we also have $t_{1i} \geq 0$, $i > 1$, so that $t_{1i}^\lambda = (1 - \lambda) r_{1i} + \lambda t_{1i} > 0$, $i > 1$, unless $\lambda = 1$ (and $t_{1i} = 0$). This shows that $T_{\lambda_0} = T$, and thus $t_1 = 0$.

Now if we add the condition $s_1 > 0$ to (8) and (9), we would restrict the space to points of $\mathcal{R}$ only. Our final result is then the following:

Any symmetric $n \times n$ matrix S which satisfies the inequalities

$$s_1 > 0 \, ,$$
$$s_{1j} \geq 0 \, , \quad j = 2, \ldots, n \, ,$$
$$S[w^{(i)}] \geq s_i \, , \quad w^{(i)} \in \mathcal{W}^{(i)} \, , \quad i = 1, \ldots, n \, ,$$

is a reduced matrix.

§5. Fundamental region[*] of the space of all matrices

Let $\mathfrak{A}$ represent the space of all $n \times n$ real, non-singular matrices. We proved in Lecture XI that if S is a symmetric, positive-definite matrix, then we can find a non-singular matrix A, such that $S = A'A$. We say that the point A of $\mathfrak{A}$ *projects* into the point S of $\mathcal{S}$. Note that, in general, there is an infinity of points A which project into the same point S.

Consider the space of all points A in $\mathfrak{A}$ which project into a point of $\mathcal{R}$. We denote this space by $\mathcal{F}_1$. If we replace $S \in \mathcal{R}$ by $S[U]$, then A is replaced by AU, and the space $\mathcal{F}_1$ by the space $\mathcal{F}_1 U$. Since the set of images $\{\mathcal{R}_U\}$ covers all of $\mathcal{P}$, as U runs over all unimodular matrices, we conclude that *the set of spaces $\{\mathcal{F}_1 U\}$ covers the whole space $\mathfrak{A}$, with no gaps or overlapping,* if *we identify U and $-U$* [overlapping on the boundary is allowed]. This is necessary because $S[-U] = S[U]$, while $-AU \neq AU$, so that $\mathcal{F}_1 = -\mathcal{F}_1$. This shows that as U runs over all unimodular matrices, the set of spaces $\{\mathcal{F}_1 U\}$ covers everything twice.

We wish to find a *fundamental region $\mathcal{F}_2$* in $\mathfrak{A}$, such that the set of regions $\{\mathcal{F}_2 U\}$ will cover everything only once, with no gaps or overlapping.

Note that

$$s_{ij} = a^{(i)} \cdot a^{(j)} \, ,$$

where $a^{(k)}$ is the kth column of A, so that the conditions for a point to lie in $\mathcal{R}$ are homogeneous quadratic inequalities in the space $\mathfrak{A}$. The boundary of $\mathcal{F}_1$ is therefore composed of surfaces of the second degree.

---

[*] One speaks nowadays of a fundamental region of a group acting on a space

We shall make $\mathcal{F}_1$ fundamental by adding the condition that $a_{11} > 0$, if $a_{11} \neq 0$; or else $a_{12} > 0$, if $a_{12} \neq 0$; $\ldots$; or else $a_{1,n-1} > 0$, if $a_{1,n-1} \neq 0$; or else $a_{1n} > 0$. Denote this region by $\mathcal{F}_2$. Then $\mathcal{F}_2$ is exactly "$\mathcal{F}_1/2$", and the sets $\mathcal{F}_2 U$, where U runs over all unimodular matrices, cover the whole space once, with no gaps and no overlapping.

For any matrix $A \in \mathfrak{A}$, there exists a unimodular matrix U, such that AU lies in $\mathcal{F}_2$, or on the boundary of $\mathcal{F}_2$. If AU lies in the interior of $\mathcal{F}_2$, U is unique. If AU is on the boundary, since $\mathcal{F}_2$ has only a finite number of neighbours [as a consequence of Lemma 4 of Lecture XIII], we could get rid of the ambiguities on the boundary and so get a space $\mathcal{F}_3$, such that $\mathcal{F}_3 U$ covers every point once and only once.

Lecture XV

§1. Volume of a fundamental region

We are interested in the volume of the fundamental region $\mathcal{F}_3$ of the space $\mathfrak{A}$ [Lecture XIV, § 5]. But if A belongs to $\mathcal{F}_3$, so does λA for $\lambda > 0$. This shows that the volume of $\mathcal{F}_3$ is infinite.

Consider, however, that part of the space $\mathfrak{A}$, for which

$$(1) \qquad\qquad |\det A| \le 1 .$$

We shall denote it by $\overline{\mathfrak{A}}$, the corresponding part of $\mathcal{F}_3$ by $\mathcal{F}$, and *its* volume by V_n. By V_n we mean the volume in a Euclidean n^2-dimensional space, where

$$(2) \qquad dV = \prod_{i,j=1}^{n} da_{ij} , \quad A = \begin{pmatrix} a_{11} & \cdots & a_{1n} \\ \vdots & & \vdots \\ a_{n1} & \cdots & a_{nn} \end{pmatrix} .$$

We shall prove that

$$(3) \qquad\qquad V_n = \int_{\mathcal{F}} dV = \frac{\zeta(2) \cdot \zeta(3) \cdot \ldots \cdot \zeta(n)}{n} ,$$

where

$$(4) \qquad\qquad \zeta(s) = \frac{1}{1^s} + \frac{1}{2^s} + \ldots , \quad s > 1 .$$

The case $n = 1$ is trivial, for $\mathcal{F}$ reduces to the interval $(0, 1]$ and then $V_1 = 1$. We shall prove the general case by induction on n. The method of proof is similar to the one used by Gauss and Dirichlet in determining the class number of quadratic fields.

§2. Outline of the proof

Let $f : \mathbb{R}^n \to \mathbb{R}$ be any Riemann integrable, bounded function which is zero for all points sufficiently far from the origin. Consider the integral

$$I = \int_{\mathbb{R}^n} f(x)\, dx \ ,$$

where dx represents the volume element in n-dimensional space. This integral can be expressed as the limit of a Riemann sum

(5) $$I = \lim_{\lambda \downarrow 0} \lambda^n \sum_g f(\lambda g) \ ,$$

where the summation is over all g-points. Note that λ^n is the volume of an n-dimensional rectangular parallelepiped [a cube, actually] containing the point λg.

Suppose we take an arbitrary matrix $A \in \mathfrak{A}$, of determinant plus or minus one. This matrix defines the basis for a lattice whose points are determined by Ag. Instead of dividing the n-dimensional space into rectangular parallelepipeds, we can divide it into parallelepipeds determined by the points Ag, and get an expression for I similar to (5). We have

(6) $$I = \lim_{\lambda \downarrow 0} \lambda^n \sum_g f(\lambda Ag) \ .$$

If the determinant of A is not necessarily plus or minus one, suppose that

$$|\det A|^{1/n} = \delta \ .$$

Then $\delta^{-1}A$ has the determinant plus or minus one, and we can apply (6). We get

(7) $$I = \lim_{\lambda \downarrow 0} \lambda^n \sum_{g \neq 0} f(\lambda \delta^{-1} Ag) \ .$$

For convenience in the following discussions, we have omitted the term for which $g = 0$. This is permissible, since $f(0)$ is bounded, and the factor λ^n will go to zero in the limit.

We now integrate (7) over the space $\mathcal{F}$. The variables will be the elements of the matrix A. We have

(8) $$I \cdot V_n = \int_{\mathcal{F}} \lim_{\lambda \downarrow 0} \lambda^n \sum_{g \neq 0} f\left(\lambda \delta^{-1} Ag\right) dV \ .$$

It can be shown [see Siegel, Annals of Math. 45 (1944) 577–622] that the interchange of integration, taking the limit, and summation, is legitimate. We shall prove that the sum

(9) $$\xi = \lambda^n \sum_{g \neq 0} \int_{\mathcal{F}} f\left(\lambda \delta^{-1} Ag\right) dV \qquad (\lambda > 0)$$

is independent of λ, and then by choosing a suitable value for λ, we shall find a relation between V_n and V_{n-1}, which will lead to the desired result (3).

§3. Change of variable

Let $g_1, \ldots, g_n$ be the components of the integral vector g, and suppose that r is the greatest common divisor of the components $g_1, \ldots, g_n$. We may split up the sum for ξ as follows:

$$(10) \qquad \xi = \xi_1 + \xi_2 + \cdots ,$$

where

$$(11) \qquad \xi_k = \lambda^n \sum_{r=k} \int_{\mathcal{F}} f\left(\lambda \delta^{-1} A g\right) dV , \qquad k = 1, 2, \ldots .$$

Note that in (11) the sum is over all integral vectors g, such that the greatest common divisor of the components is k. In particular, we have

$$(12) \qquad \lambda^{-n} \xi_1 = \sum_{g}{}' \int_{\mathcal{F}} f\left(\lambda \delta^{-1} A g\right) dV ,$$

where the sum is extended over all primitive g's.

We have proved before that any primitive vector g may be filled up to a unimodular matrix U_g. Note that

$$(13) \qquad U_g^{-1} g = e^{(1)} ,$$

the vector whose first component is one, while all the other components are zero. [For each primitive g, we choose, once and for all, such a matrix U_g.]

We may write A as $A U_g U_g^{-1}$, and use $A U_g$ as the new variable of integration. The new domain of integration will be $\mathcal{F} U_g$. Since each row of A, and the corresponding set of n variables, is transformed by a unimodular matrix, the Jacobian of the transformation will be 1, and we get

$$(14) \qquad \lambda^{-n} \xi_1 = \sum_{g}{}' \int_{\mathcal{F} U_g} f\left(\lambda \delta^{-1} x\right) dV ,$$

where x is the first column of the matrix $A U_g$ with $A \in \mathcal{F}$. The integrand is independent of g, since x runs over the first column of any matrix which is in the space $\mathcal{F} U_g$. Since $\mathcal{F} U_{g_1} \cap \mathcal{F} U_{g_2} = \emptyset$ for $g_1 \neq g_2$, we can carry out the summation over the domain of integration, and obtain

$$(15) \qquad \lambda^{-n} \xi_1 = \int_{\mathcal{F}_1} f\left(\lambda \delta^{-1} x\right) dV ,$$

where [*not to be confused with the $\mathcal{F}_1$, $\mathcal{F}_2$ in Lecture XIV*]

$$(16) \qquad \mathcal{F}_1 = \bigcup_{g \text{ primitive}} \mathcal{F} U_g .$$

We shall prove that $\mathcal{F}_1$ is the fundamental region of $\overline{\mathfrak{A}}$ under a certain subgroup of the unimodular group.

Note that if U_g is multiplied by a matrix of the form

(17)
$$\begin{pmatrix} 1 & u \\ 0 & U^* \end{pmatrix},$$

where U^* is a unimodular matrix of $n-1$ rows and columns, and u is a row of integers, the result will be again a unimodular matrix U, whose first column is g. The converse is also true. Given any unimodular matrix U, with g as its first column, there exists a unique matrix of the form (17), such that

(18)
$$U = U_g \begin{pmatrix} 1 & u \\ 0 & U^* \end{pmatrix}.$$

The proof is a trivial consequence of (13), and the fact that U_g^{-1} is unimodular.

Now let U be any unimodular matrix. Since all its columns must be primitive, and in particular its first, it follows that we can write U uniquely as in (18).

Note that the matrices of the form (17) form a subgroup $\mathfrak{G}$ of the unimodular group.

We know that $\mathcal{F}U$, where U runs over all unimodular matrices, covers $\overline{\mathfrak{A}}$ once and once only, but by (18),

$$\bigcup_{\substack{U \\ \text{unimodular}}} \mathcal{F}U = \bigcup_{\begin{pmatrix} 1 & u \\ 0 & U^* \end{pmatrix} \in \mathfrak{G}} \bigcup_{g \text{ primitive}} \mathcal{F}U_g \begin{pmatrix} 1 & u \\ 0 & U^* \end{pmatrix} = \mathcal{F}_1 \bigcup_{\begin{pmatrix} 1 & u \\ 0 & U^* \end{pmatrix} \in \mathfrak{G}} \begin{pmatrix} 1 & u \\ 0 & U^* \end{pmatrix}.$$

This shows that $\mathcal{F}_1$ is a fundamental region of $\overline{\mathfrak{A}}$ with reference to the subgroup $\mathfrak{G}$ of the unimodular group.

§4. A new fundamental region

We are going to find a new fundamental region for $\overline{\mathfrak{A}}$, but first we shall obtain an expression for the most general matrix in $\overline{\mathfrak{A}}$, with a given first column.

Denote the given column by x, and suppose it is completed to a real matrix A_x of determinant 1; for example, if $x_1 \neq 0$, as follows:

(19)
$$A_x = \begin{pmatrix} x_1 & 0 & 0 & \cdots & 0 \\ x_2 & x_1^{-1} & 0 & \cdots & 0 \\ x_3 & 0 & 1 & \cdots & 0 \\ \vdots & \vdots & \vdots & \ddots & \vdots \\ x_n & 0 & 0 & \cdots & 1 \end{pmatrix}.$$

[The matrices in $\overline{\mathfrak{A}}$ which have their first diagonal element equal to zero are of volume zero in $\overline{\mathfrak{A}}$. Therefore we will neglect them in the following discussion.] Then, just as before, it is easy to see that if A has x as its first column, then

$$(20) \qquad A = A_x \begin{pmatrix} 1 & a \\ 0 & A^* \end{pmatrix} ,$$

where A^* is a $(n-1) \times (n-1)$ matrix whose determinant equals det A.

We now introduce the elements of x, a, and A^* as the new variables of integration. If we denote the elements of a by a_k, $k = 2, \ldots, n$, and the elements of A^* by a_{ik}^*, $(i, k = 2, \ldots, n)$, we find from (19) and (20) that

$$\begin{aligned}
a_{ik} &= x_i a_k + a_{ik}^* & (i \geq 3, \quad k \geq 2) \\
a_{2k} &= x_2 a_k + x_1^{-1} a_{2k}^* & (k \geq 2) \\
a_{1k} &= x_1 a_k & (k \geq 2) \\
a_{i1} &= x_i & (i \geq 1) .
\end{aligned}$$

It is clear from these equations that the transformation from a_{ik} to a_{ik}^* $(i \geq 3, \; k \geq 2)$ will leave the volume element unchanged, the transformation from a_{2k} to a_{2k}^* $(k \geq 2)$ will multiply the volume element by $x_1^{-(n-1)}$, while the transformation from a_{1k} to a_k $(k \geq 2)$ will multiply it by x_1^{n-1} and the transformation from a_{i1} to x_i $(i \geq 1)$ will leave it unchanged. The total effect of the complete transformation will be to leave the volume element unchanged, so that we may write

$$dV = dx \, da \, dV^* .$$

Consider now any element V of the group $\mathfrak{G}$ defined previously. We know by (20) and (17) that

$$(21) \qquad AV = A_x \begin{pmatrix} 1 & a \\ 0 & A^* \end{pmatrix} \begin{pmatrix} 1 & u \\ 0 & U^* \end{pmatrix} = A_x \begin{pmatrix} 1 & u + aU^* \\ 0 & A^*U^* \end{pmatrix} .$$

Since U^* is an arbitrary $(n-1) \times (n-1)$ unimodular matrix, it is always possible to determine U^* so that A^*U^* will lie in $\mathcal{F}^*$, the fundamental region in $\overline{\mathfrak{A}}$ for $(n-1) \times (n-1)$ matrices under unimodular transformations. We then determine u so that all elements of $u + aU^*$ lie between zero and one, zero included but one excluded. Since the elements of u are integers, it follows that V is uniquely determined.

We define $\mathcal{F}_2$ as the region containing all matrices of the form

$$A_x \begin{pmatrix} 1 & a \\ 0 & A^* \end{pmatrix} ,$$

where the elements of a are between zero and one, and A^* lies in $\mathcal{F}^*$. We have just shown that given any matrix $A \in \overline{\mathfrak{A}}$, there exists a unique matrix V in $\mathfrak{G}$, such that AV lies in $\mathcal{F}_2$; therefore $\mathcal{F}_2$ is also a fundamental region of $\overline{\mathfrak{A}}$ under the group $\mathfrak{G}$.

§5. Integrals over fundamental regions are equal

We shall prove that

$$(22) \qquad \int_{\mathcal{F}_1} f\left(\lambda \delta^{-1} x\right) dV = \int_{\mathcal{F}_2} f\left(\lambda \delta^{-1} x\right) dV .$$

Note that the integrand is invariant under the group $\mathfrak{G}$, where $\mathfrak{G}$ acts from the right on the argument of f, since multiplication by any element of $\mathfrak{G}$ leaves the first column of a matrix unchanged.

Since all the matrices belonging to $\mathfrak{G}$ are unimodular and so contain only integral elements, it follows that $\mathfrak{G}$ is countable. We denote the elements of $\mathfrak{G}$ by $V_1, V_2, \ldots$.

Consider the union

$$\bigcup_i \mathcal{F}_2 V_i .$$

We know that $\mathcal{F}_2 V_i$, $i = 1, 2, \ldots$, cover the whole space $\overline{\mathfrak{A}}$ without gaps or overlapping. Therefore the union must cover the region $\mathcal{F}_1$. Let $\mathcal{G}_k$ be the intersection of $\mathcal{F}_1$ and $\mathcal{F}_2 V_k$. It is clear that the sets $\mathcal{G}_k$ have no interior points in common, and that

$$(23) \qquad \mathcal{F}_1 = \bigcup_k \mathcal{G}_k .$$

Note also that the intersection of $\mathcal{F}_1 V_k^{-1}$ and $\mathcal{F}_2$ is the space $\mathcal{G}_k V_k^{-1}$. Since V_k^{-1} runs over all elements of $\mathfrak{G}$ and since $\mathcal{F}_1$ is a fundamental region of the space $\overline{\mathfrak{A}}$ with respect to $\mathfrak{G}$, it follows that the union

$$\bigcup_k \mathcal{F}_1 V_k^{-1}$$

covers $\mathcal{F}_2$, so that we have

$$(24) \qquad \bigcup_k \mathcal{G}_k V_k^{-1} = \mathcal{F}_2 .$$

We can now complete the proof of (22). On using (23), (24), and the fact that the integrand is invariant under the group $\mathfrak{G}$ we have

$$\int_{\mathcal{F}_1} = \int_{\bigcup_k \mathcal{G}_k} = \int_{\bigcup_k \mathcal{G}_k V_k^{-1}} = \int_{\mathcal{F}_2} .$$

§6. Evaluation of the integral

From (15) and (22) we know that

$$(25) \qquad \lambda^{-n} \xi_1 = \int_{\mathcal{F}_1} f\left(\lambda \delta^{-1} x\right) dV = \int_{\mathcal{F}_2} f\left(\lambda \delta^{-1} x\right) dx \, da \, dV^* .$$

In $\mathcal{F}_2$, a runs over the unit cube in $(n-1)$ dimensions, A^* [corresponding to dV^*, see (2)] lies in the fundamental region $\mathcal{F}^*$ for $\overline{\mathfrak{A}}$ in $(n-1)$ dimensions with respect to the unimodular group, and x runs over the entire n-dimensional Euclidean space $\mathbb{R}^n$ except for a set of volume zero.

If we note that after replacing x by $\delta\lambda^{-1}x$,

$$\int_{\mathbb{R}^n} f\left(\lambda\delta^{-1}x\right) dx = I \cdot \left(\frac{\delta}{\lambda}\right)^n ,$$

we may then write (25) as follows:

$$\lambda^{-n}\xi_1 = \int_{\mathcal{F}^*}\int_{\mathbb{R}^n} f\left(\lambda\delta^{-1}x\right) dx\, dV^* = \int_{\mathcal{F}^*} I \cdot \left(\frac{\delta}{\lambda}\right)^n dV^*$$

$$= \lambda^{-n}I \int_{\mathcal{F}^*} |\det A|\, dV^* = \lambda^{-n}I \int_{\mathcal{F}^*} |\det A^*|\, dV^* ,$$

so that

(26) $$\xi_1 = I \int_{\mathcal{F}^*} |\det A^*|\, dV^* .$$

By definition, we have

$$V_n = \int_{\mathcal{F}} dV ,$$

and if we multiply all matrix elements by $\kappa^{1/n}$, $0 \leq \kappa \leq 1$, we get

$$\int_{\mathcal{F}_\kappa} dV = V_n \kappa^n ,$$

where $\mathcal{F}_\kappa$ is the intersection of the region $\mathcal{F}$ and the region for which $|\det A| \leq \kappa$. Now

(27) $$\int_{\mathcal{F}} |\det A|\, dV = \int_0^1 \kappa \frac{\partial}{\partial\kappa} \int_{\mathcal{F}_\kappa} dV\, d\kappa = \int_0^1 \kappa V_n n \kappa^{n-1} d\kappa = \frac{n}{n+1} \cdot V_n .$$

Using (27) in (26) we get

(28) $$\xi_1 = I \cdot \frac{n-1}{n} \cdot V_{n-1} .$$

Similarly in ξ_k, since each non-zero g-vector is k times a primitive g-vector, the sum for ξ_k, that is,

$$\xi_k = \lambda^n \sum_{\text{g.c.d.}\{g_1,\ldots,g_n\}=k} \int_{\mathcal{F}} f\left(\lambda\delta^{-1}Ag\right) dV$$

$$= \left(\frac{\lambda}{k}\right)^n \sum_g{}' \int_{\mathcal{F}} f\left(\lambda\delta^{-1}Ag\right) dV ,$$

reduces to

$$\xi_k = \xi_1 k^{-n} \ .$$

From this it follows that

(29)
$$\xi = \sum_{k=1}^{\infty} \xi_k = \xi_1 \cdot \zeta(n) \ ,$$

where

$$\zeta(n) = 1^{-n} + 2^{-n} + \dots \ , \quad n > 1 \ .$$

Since ξ is independent of λ, we may take $\lambda = 1$ in (9), and get

(30)
$$I \cdot V_n = \xi = \xi_1 \cdot \zeta(n) = I \cdot \frac{n-1}{n} \cdot V_{n-1} \cdot \zeta(n) \ ,$$

from (28).

By induction on (30), and on condition that $I \neq 0$ (which can be secured by suitable choice of f), we find that

$$nV_n = (n-1)V_{n-1} \cdot \zeta(n) = \zeta(2) \cdot \zeta(3) \cdot \dots \cdot \zeta(n) \ , \quad (\text{since } V_1 = 1)$$

so that we have finally proved that

$$V_n = \frac{\zeta(2) \cdot \zeta(3) \cdot \dots \cdot \zeta(n)}{n} \ , \quad n > 1 \ .$$

§7. Generalizations of Minkowski's First Theorem

Consider in $\mathbb{R}^n$ any convex body $\mathcal{B}$ with the origin as centre. Minkowski's First Theorem states that for any lattice of determinant 1, there exists a lattice point different from the origin in $\overline{\mathcal{B}}$, if the volume of $\mathcal{B}$ is 2^n. This value 2^n is the best possible constant for the set of all convex bodies. For more specialized sets, this may not be the best constant. We shall discuss several particular cases.

Let C be an arbitrary measurable, bounded set in $\mathbb{R}^n$, and suppose that for every lattice of determinant 1, there exists a non-zero lattice point in C. Then we shall prove that the volume of C is equal to or greater than 1.

Let f be the characteristic function of C, that is,

$$f(x) = \begin{cases} 1 \ , & \text{for } x \in C \ , \\ 0 \ , & \text{for } x \notin C \ ; \end{cases}$$

so that

$$I = \int_{\mathbb{R}^n} f(x) \, dx = \text{volume of } C \ .$$

The vectors $\delta^{-1}Ag$ run over all non-zero vectors of a lattice of determinant 1, if g runs over all non-zero g-points in $\mathbb{R}^n$, and therefore, by hypothesis,

$$\sum_{g \neq 0} f\left(\delta^{-1}Ag\right) \geq 1 .$$

From this it follows that

$$I \cdot V_n \geq \int_{\mathcal{F}} dV = V_n ,$$

so that

(31) $I \geq 1 .$

This result is due to E. Hlawka [Math. Zeit. *49* (1943/44) 285–312].

Suppose now that C is a *star domain* in $\mathbb{R}^n$, $n \geq 2$; that is to say, if x lies in C, then all λx lie in C, where $0 \leq \lambda \leq 1$. Suppose again that C contains at least one non-zero lattice point of every lattice with determinant 1. Then we shall prove that

(32) $I \geq \zeta(n) .$

If C is a star domain, it must contain at least one primitive vector, since with g also $\frac{1}{r}g$ belongs to C for $r \geq 1$; therefore

$$\sum_{g}' f\left(\delta^{-1}Ag\right) \geq 1 ,$$

and using (29), we find that

$$I \cdot V_n \geq \zeta(n) \int_{\mathcal{F}} dV = \zeta(n) V_n ,$$

so that

$$I \geq \zeta(n) .$$

Suppose now that the star domain has 0 as centre. Then we shall prove that

(33) $I \geq 2 \cdot \zeta(n) .$

This follows easily from the above reasoning. If $\delta^{-1}Ag$ belongs to C, then so does $-\delta^{-1}Ag$, and we have

$$\sum_{g}' f\left(\delta^{-1}Ag\right) \geq 2 .$$

These theorems about star domains, and symmetrical star domains, were conjectured by Minkowski, and first proved by Hlawka. [For later work see Schmidt and other references cited at the end.]

§8. A lower bound for the packing of spheres

Let $\mathcal{C}$ be a sphere in $\mathbb{R}^n$, $n \geq 2$, of radius r, so that

$$I = \sigma_n r^n \, ,$$

where $\sigma_n = \frac{\pi^{n/2}}{\Gamma(\frac{n}{2}+1)}$. Consider all lattices of determinant 1. For each such lattice, which is defined by A, we are interested in the lattice point, different from the origin, lying nearest to the origin. If we call λ the distance of this point from the origin, we have

$$\lambda = \min_{g \in \mathbb{Z}^n - 0} |Ag| \, ,$$

and

$$\lambda^2 = \min_{g \in \mathbb{Z}^n - 0} g'A'Ag = \min_{g \in \mathbb{Z}^n - 0} S[g] \, ,$$

where $S = A'A$, so that λ^2 is the minimum value of $S[x]$ on the lattice of all g-points excluding the origin.

Consider all symmetric, positive-definite matrices S such that $\det S = 1$. For any S, there exists a minimum value on the lattice of g-points excluding the origin. Call the supremum of these minimum values γ_n. By Minkowski's First Theorem,

$$\sigma_n \gamma_n^{n/2} \leq 2^n \, .$$

Since the sphere is a star domain with centre, we may apply (33) to it, and get

(34) $$\sigma_n \gamma_n^{n/2} \geq 2 \cdot \zeta(n) \, .$$

We showed previously [see Lecture XII] that if q_n represents the ratio of covered space to the whole space, when spheres are packed closest together, then

$$q_n = \frac{\sigma_n \cdot \gamma_n^{n/2}}{2^n} \, .$$

Using (34), we finally get

$$q_n \geq \frac{\zeta(n)}{2^{n-1}} \, , \quad (n \geq 2) \, .$$

References

Siegel's references

Blichfeldt, H.F.:
- (i) A new principle in the geometry of numbers, with some applications. Trans. Amer. Math. Soc. *15* (1914) 227–235
- (ii) The minimum value of quadratic forms and the closest packing of spheres. Math. Annalen *101* (1929) 605–608

Bôcher, M.: Introduction to higher algebra. New York Macmillan 1907

Davenport, H.: Note on the product of three homogeneous linear forms. J. London Math. Soc. *16* (1941) 98–101

Gauss, C.F.: Besprechung des Buches von L.A. Seeber: Untersuchungen über die Eigenschaften der positiven ternären quadratischen Formen. Göttingische gelehrte Anzeigen (1831); Werke II, 188–196

Hadamard, J.: Résolution d'une question relative aux déterminants. Bull. Sci. Math. 2e série *17* (1893) 240–246; Œuvres I, 239–245

Hajós, G.: Über einfache und mehrfache Bedeckung des n-dimensionalen Raumes mit einem Würfelgitter. Math. Zeit. *47* (1942) 427–467

Hlawka, E.: Zur Geometrie der Zahlen. Math. Zeit. *49* (1943/44) 285–312

Hurwitz, A.: Über die angenäherte Darstellung der Irrationalzahlen durch rationale Brüche. Math. Annalen *39* (1891) 279–284; Werke II, 122–128

Kronecker, L.: Näherungsweise ganzzahlige Auflösung linearer Gleichungen. Berliner Sitzungsberichte (1894); Werke III (i), 47–109

Mahler, K.: On Minkowski's theory of reduction of positive definite quadratic forms. Quarterly J. Math. Oxford *9* (1938) 259–262

Minkowski, H.:
- (i) Geometrie der Zahlen. Teubner 1896
- (ii) Diophantische Approximationen. Teubner 1907
- (iii) Gesammelte Abhandlungen I, II.

Mordell, L.J.: Observation on the minimum of a positive quadratic form in eight variables. J. London Math. Soc. *19* (1944) 3–6

Pontrjagin, L.: Topological groups. (1939) Princeton [Th. 32, p. 134]

Siegel, C.L.: On the theory of indefinite quadratic forms. Annals of Math. *45* (1944) 577–622; Ges. Abh. II, 421–466

Weyl, H.: On geometry of numbers. Proc. London Math. Soc. (2) *47* (1942) 268–289; Ges. Abh. IV, 75–96

Additional references

Cassels, J.W.S.: An introduction to the geometry of numbers. Second printing, corrected (1971), Springer-Verlag

Conway, J.H., and Sloane, N.J.A.: Sphere packings, lattices and groups. Springer-Verlag 1988

Gruber, P.M., and Lekkerkerker, C.G.: Geometry of numbers. North-Holland 1987

Hlawka, E.:

   (i) Theorie der Gleichverteilung. Bibliographisches Institut 1979

   (ii) 90 Jahre Geometrie der Zahlen. Jahrbuch Überblicke Mathematik (1980) 9–41

   (iii) Selecta. Springer-Verlag (to appear)

Pontryagin, L.S.: Topological groups. Second edition (1966), Gordon and Breach [Th. 52, p. 273]

Schmidt, W.M.:

   (i) Eine Verschärfung des Satzes von Minkowski-Hlawka. Monatshefte für Mathematik *60* (1956) 110–113

   (ii) Masstheorie in der Geometrie der Zahlen. Acta Math. *102* (1959) 159–224

   (iii) On the Minkowski-Hlawka theorem. Illinois J. Math. *7* (1963) 18–23; 714

Siegel, C.L.:

   (i) Neuer Beweis des Satzes von Minkowski über lineare Formen. Math. Annalen *87* (1922) 36–38; Ges. Abh. I, 165–167; and IV, 339

   (ii) Über Gitterpunkte in convexen Körpern und ein damit zusammenhängendes Extremalproblem. Acta Math. *65* (1935) 307–323; Ges. Abh. I, 311–325

   (iii) A mean value theorem in geometry of numbers. Annals of Math. *46* (1945) 340–347; Ges. Abh. III, 39–46

Verner, L.: Quadratic forms in an adelic setting. L'Enseignement Math. *23* (1977) 7–12

Weil, A.: Sur quelques résultats de Siegel. Summa Brasiliensis Mathematicae *1* (1946) 21–39; Collected Papers I, 339–357

Index

C. L. Siegel

Gesammelte Abhandlungen

Herausgeber: K. Chandrasekharan, H. Maaß

3. Bände, die nur zusammen abgegeben werden

1966. 1 Portrait, XVI, 1523 Seiten (381 Seiten in
Englisch, 21 Seiten in Französisch).
ISBN 3-540-03658-X

Carl Ludwig Siegel hat mit seinem tiefgründigen
und ausgereiften Werk die Entwicklung der
Mathematik vor allem auf den Gebieten der
komplexen Analysis und der Zahlentheorie
wesentlich beeinflußt. Die Veröffentlichung der in
mehr als 30 Zeitschriften und Festbänden publi-
zierten Arbeiten ist für jeden, der auf diesen
Gebieten, aber auch in der algebraischen Geometrie, der Himmelsmechanik und der Zahlentheorie
arbeitet, eine Quelle wichtiger Anregungen. Unter
den abgedruckten Aufsätzen befindet sich auch
die grundlegende Monographie ‚Zur Reduktionstheorie quadratischer Formen'. Der Aufsatz ‚Zur
Geschichte des Frankfurter Mathematischen
Seminars' hebt sich als einziger nicht-mathema-
tischer Beitrag von dem Gesamtwerk ab; er ist vor
allem dem Andenken der Mathematiker Dehn,
Epstein, Hellinger und Szász gewidmet. Ein Fak-
simileabdruck eines handschriftlichen Briefes
Siegels an W. Gröbner gibt Zeugnis von einer
gewissenhaften Art schriftlicher Mitteilung, wie sie
heute nur noch selten gepflegt wird.

Springer-Verlag Berlin
Heidelberg New York London
Paris Tokyo Hong Kong

Springer

Band 4

1979. 1 Portrait. V, 343 Seiten. ISBN 3-540-09374-5

C. L. Siegel hat mit seinen Publikationen über mehr als fünf Jahrzehnte nachhaltig auf die Entwicklung der Mathematik Einfluß genommen.

Dieser vierte Band seiner **Gesammelten Abhandlungen** enthält sein Alterswerk, die nach 1968 erschienenen Veröffentlichungen. Er enthält wichtige Beiträge zu Gebieten, auf deren Entwicklung Siegel maßgeblichen Einfluß hatte: Berechnung von Zetafunktionen an ganzzahligen Stellen, Fouriersche Koeffizienten von Modulformen, Modulen Abelscher Funktionen, Theorie der quadratischen Formen. Auch enthalten ist die Abhandlung „Zu den Beweisen des Vorbereitungssatzes von Weierstraß", mit Siegels eigenem schönen Beweis dieses Satzes.

Inhalt: Zu den Beweisen des Vorbereitungssatzes von Weierstraß. – Bernoullische Polynome und quadratische Zahlkörper. – Zum Beweise des Starkschen Satzes. – Über die Fourierschen Koeffizienten von Eisensteinschen Reihen der Stufe T. – Erinnerungen an Frobenius. – Abschätzung von Einheiten. – Berechnung von Zetafunktionen an ganzzahligen Stellen. – Über die Fourierschen Koeffizienten von Modulformen. – Einige Erläuterungen zu Thues Untersuchungen über Annäherungswerte algebraischer Zahlen und diophantische Gleichungen. – Algebraische Abhängigkeit von Wurzeln. – Über Moduln Abelscher Funktionen. – Periodische Lösungen von Differentialgleichungen. – Wurzeln Heckescher Zetafunktionen. – Zur Theorie der quadratischen Formen. – Normen algebraischer Zahlen. – Beitrag zum Problem der Stabilität. – Zur Summation von L-Reihen. – Vorwort „Zur Reduktionstheorie quadratischer Formen". – Vollständige Liste aller Titel (Bd. I–Bd. IV). – Titel aller Bücher und Vorlesungsausarbeitungen. – Berichtigungen und Bemerkungen, Bd. I–Bd. III betreffend. – Nachwort.

Springer-Verlag Berlin
Heidelberg New York London
Paris Tokyo Hong Kong

Springer